T0331891

A Quantum Leap in Information Theory

A Quantum Leap in Information Theory

Stefano Mancini
University of Camerino, Italy

Andreas Winter
ICREA and Autonomous University of Barcelona, Spain

NEW JERSEY • LONDON • SINGAPORE • BEIJING • SHANGHAI • HONG KONG • TAIPEI • CHENNAI • TOKYO

Published by

World Scientific Publishing Co. Pte. Ltd.
5 Toh Tuck Link, Singapore 596224
USA office: 27 Warren Street, Suite 401-402, Hackensack, NJ 07601
UK office: 57 Shelton Street, Covent Garden, London WC2H 9HE

British Library Cataloguing-in-Publication Data
A catalogue record for this book is available from the British Library.

The cover design, suggesting binary symbols encoded into strangely shaped quantum systems, is borrowed from the series of works "I quanta" (1959–1960) by Lucio Fontana.

A QUANTUM LEAP IN INFORMATION THEORY

ISBN 978-981-120-154-7 (hardcover)
ISBN 978-981-120-155-4 (ebook for institutions)
ISBN 978-981-120-156-1 (ebook for individuals)

For any available supplementary material, please visit
https://www.worldscientific.com/worldscibooks/10.1142/11314#t=suppl

Desk Editor: Nur Syarfeena Binte Mohd Fauzi

Typeset by Stallion Press
Email: enquiries@stallionpress.com

Printed in Singapore

CONTENTS

PREFACE

About 70 years after the foundation of information theory by Claude E. Shannon, it is being realized that it has undergone a leap with the development of a quantum theory of information. The latter is by its very nature the product of a multidisciplinary field of research. In recent years, more and more students from different backgrounds are approaching it. At the same time, as notions and methods developed in this field are spreading across other branches of physics, mathematics, and computer science, quantum information theory is no longer viewed as a specialized subject suitable for an advanced course, but rather as a fundamental one that can be taught in a basic interdisciplinary course.

The present book provides a self-contained introduction to the field of quantum information theory to newcomers at upper undergraduate or graduate level. Notions and properties of quantum information theory are presented in a pedagogical way by also referring to the classical theory of information and stressing differences and similarities between the classical and the quantum frameworks.

The fundamental link between physics and information theory is provided by the possibility to consider an alphabet's symbol as representative of classical physical states. When a state is not exactly known, the notion of probability distribution on the alphabet naturally emerges. This in turn leads to the notion of random variable and its entropy. Furthermore, while closed systems imply deterministic state changes, open systems give rise to stochastic state changes leading to the notion of channel maps. Then this picture will be discussed and transposed to the quantum framework step by step.

Specifically, the book is structured as follows. After a prelude introducing the notions of alphabets and classical spaces of states, the Shannon entropy, as well as other related entropic quantities and their properties, will be discussed. After that, the postulates of quantum mechanics are presented, followed by a discussion on how the quantum framework subsumes the classical one. Then the notions and significance of the von Neumann entropy and other quantum entropic functionals will be introduced. Moving on, stochastic maps in both the classical and quantum settings are described. An interlude will be devoted to estimation theory focusing on channel maps. The core part of the book follows. First, protocols of data compression are described. Next data transmission, where upper bounds for data transmission rates are derived. At this point, error correcting codes will be presented as a method to counteract the effect of noise. Having all the tools at hand, the issue of channel capacities is addressed formulating the coding theorems and showing that the upper bounds for data transmission rates previously presented (converse part) can be actually achieved (direct part). After that the manipulation of a purely quantum resource, namely entanglement, will be described. The final chapter will be devoted to basic notions of classical and quantum cryptography with a security proof of the latter. An appendix is devoted to revisiting fundamentals of linear algebra in the light of the Dirac formalism.

Finally, at the end of each chapter, exercises are proposed to enhance the understanding of the concepts that are introduced. Some of these exercises should be viewed as useful complements: the reader is encouraged to spend some time on them (solutions of advanced exercises are given at the end of the book).

NOTATIONS

Mathematical notations. $\mathbb{R}$ (resp. $\mathbb{C}$) is the set of all real (resp. complex) numbers. The complex conjugation of a number z is denoted by a star superscript, z^*. Furthermore, we denote by $\mathbb{N} = \{1, 2, \ldots\}$ and $\mathbb{Z} = \{\ldots, -2, -1, 0, +1, +2, \ldots\}$ the set of positive integers and the set of all integers, respectively.

Alphabets are simply finite sets of symbols, denoted by $\mathcal{A}$, $\mathcal{X}$, $\mathcal{Y}$ or $\mathcal{Z}$. Furthermore, $\mathbb{F}_q$ indicates the finite field of cardinality q, where the latter is a prime power, and usually intended as a prime.

The Shannon entropy is denoted by H, while the von Neumann entropy is S. The logarithm (log) is always intended to base 2.

$\mathcal{H}$ denotes a (finite-dimensional) Hilbert space. When necessary, $\mathcal{H}$ is supplied with a label indicating the system A, B, C, ... to which is associated. For vectors in $\mathcal{H}$ and their dual we use the Dirac notation, namely $|\psi\rangle$ and $\langle\psi|$. The symbol $\mathfrak{L}(\mathcal{H})$ stands for the set of linear operators acting on $\mathcal{H}$.

The adjoint of an operator U is indicated by $U^\dagger$. If necessary we indicate the initial and final space as $U^{A\to B}$ for $U : \mathcal{H}_A \to \mathcal{H}_B$. Unless explicitly required we do not distinguish between an operator and its matrix representation. Pauli operators are indicated by $\sigma^X, \sigma^Y, \sigma^Z$. Furthermore, Tr denotes the trace.

$\mathfrak{P}(\mathcal{X})$ and $\mathfrak{D}(\mathcal{H})$ indicate the set of probability mass functions on $\mathcal{X}$ and the set of density operators on $\mathcal{H}$, respectively.

Γ denotes a classical error correcting code and $\mathcal{C}$ a quantum error correcting code.

T indicates the set of typical sequences of $\mathcal{X}^n$, while $\mathcal{T}_\rho$ the typical subspace of $\mathcal{H}^{\otimes n}$ induced by ρ.

$C(\bullet)$ and $Q(\bullet)$ stand for the classical and the quantum capacity respectively of the channel map in the argument. Analogously $C_E(\bullet)$ and $Q_E(\bullet)$ stand for the classical and the quantum capacity assisted by entanglement.

Groups are denoted by roman fonts, e.g. $\mathrm{U}(n)$ the unitary group of a Hilbert space of dimension n, $\mathrm{SU}(n)$ the special unitary group in dimension n, $\mathrm{P}(n)$ the Pauli group in dimension n, etc.

When there is no ambiguity we use the same notation for symbols of $\mathbb{F}_q$ and for vectors on $\mathbb{F}_q^n$. Otherwise we distinguish them as x and x^n respectively. Analogously when there is no ambiguity we use the same notation for elements of $\mathbb{R}$ and $\mathbb{R}^n$. Otherwise we distinguish them as v and $\vec{v}$.

Conventions. Capital letters such as X, Y and Z are mostly used for discrete random variables; p_X is the probability distribution of X (whenever not necessary, the subscript referring to the random variable is omitted). Moreover, depending on the context, p_X is also intended as the vector whose components are given by the probability values for the possible outcomes of X. Pr denotes the probability and $\mathbb{E}$ the expectation value.

The fidelity, be it either classical or quantum, is indicated by F. For the two cases it can be distinguished thanks to the arguments.

We distinguish the identity operator on $\mathcal{H}$ from the identity map on $\mathfrak{D}(\mathcal{H})$, and denote them as I and id, respectively.

Greek letters E, Δ, N refer to classical channel maps (encoding, decoding, noisy), while calligraphic letters $\mathcal{E}$, $\mathcal{D}$, $\mathcal{N}$ refer to quantum channel maps. $\mathtt{R}$ stands for a rate (we use $\mathtt{K}$ when referring to a secret key).

Acronyms. The acronym POVM means positive operator valued measure. CPTP stands for completely positive and trace preserving. The acronym LOCC means local operations and classical communication. QKD means quantum key distribution.

Furthermore, the following standard mathematical abbreviations will be used: *r.h.s.* for "right hand side", *l.h.s.* for "left hand side", *w.l.o.g.* for "without loss of generality", *iff* for "if and only if", *pdf* for "probability distribution function" and i.i.d. for "independent and identically distributed".

Chapter 1

PRELUDE

This chapter is devoted to linking elementary notions of information to basic classical physical concepts. In such a manner, we pave the way to deeper relations that will be established in subsequent chapters, always using first a classical approach and then moving to the quantum setting.

1.1 Alphabets and spaces of states

An alphabet $\mathcal{X}$ is a set of symbols. If the latter are countable, we speak of a discrete alphabet, otherwise a continuous alphabet. We will consider henceforth only discrete alphabets. Moreover, if $|\mathcal{X}| = d < +\infty$, we speak of a finite ("d-ary") alphabet.

Let us focus on the d-ary alphabet $\mathcal{X} = \{0, 1, \ldots, d-1\}$, and introduce on it the addition modulo d denoted by $\oplus$. With respect to this operation, $\mathcal{X}$ is an abelian group. Moreover, if d is a prime, with respect to the integer multiplication modulo d, the set $\mathcal{X}\backslash\{0\}$ is an abelian group as well. Finally, it can easily be checked that the integer multiplication modulo d satisfies the distributive law with respect to $\oplus$, namely

$$a(b \oplus c) = (ab) \oplus (ac), \quad \forall a, b, c \in \mathcal{X}.$$

Hence, $\mathcal{X} = \{0, 1, \ldots, d-1\}$ results in a finite field of order d, usually denoted by $\mathbb{F}_d$ (also often denoted GF(d), because they are known as Galois fields).

The state of a physical system can be regarded as a mathematical object from which we can derive the properties of the system itself.

Classically speaking, it is identified with the values taken by relevant observables (measurable physical quantities) of the system.

Then a finite alphabet $\mathcal{X}$ can be used as a representative of the finite set of states of a (classical) physical system. Each symbol of the alphabet will denote a possible state of the system.

For $\mathcal{X} = \mathbb{F}_d$ (d prime), we are dealing with a physical system with only d possible states characterized by the d values of a single observable.

In case of a composite system, the state space will be represented by the Cartesian product of the spaces of states of subsystems. A relevant case is given by n identical subsystems, each having $\mathbb{F}_d$ as state space. Then, the state space of the composite system reads $\mathbb{F}_d^n$.

The space $\mathbb{F}_d^n$ can be regarded as a vector space over the field $\mathbb{F}_d$. Its vectors are strings $v = (v_1, \ldots, v_n)$ with $v_i \in \mathbb{F}_d$. This vector space is known as *Hamming space*. There, the vectors addition is defined as

$$u + v := (u_1 \oplus v_1, \ldots, u_n \oplus v_n),$$

and the multiplication of scalar by vector as

$$\alpha u := (\alpha u_1, \ldots, \alpha u_n), \quad \alpha \in \mathbb{F}_d.$$

In the Hamming space $\mathbb{F}_d^n$, we can introduce the following inner product:

$$u \cdot v := \bigoplus_{i=1}^{n} u_i v_i, \tag{1.1}$$

where the product of components is the intended modulus d. Notice however that such an inner product is degenerate, i.e. it can be that $u \cdot u = 0$ even if $u \neq 0$. Hence, it will not induce a proper norm and consequently a proper notion of distance.

We are then led to introduce a proper notion of distance on $\mathbb{F}_d^n$ as follows:

Definition 1.1.1. Given $u, v \in \mathbb{F}_d^n$, the Hamming distance $d(u, v)$ is the number of components by which the strings u and v differ from each other.

A *closed* physical system is characterized by the fact that state change occurs deterministically. Then, in $\mathbb{F}_d$ they can only be described by bijective applications, i.e. permutations $\pi : \mathbb{F}_d \to \mathbb{F}_d$.

Furthermore, being the state of a physical system identified by the values of relevant observables, the measurement of an observable gives a unique (with certainty) result (the one characterizing the state) and the state is left unchanged.

We can now summarize what we discussed in the following way:

- To any isolated physical system, having a finite number of possible states d, we associate a state space $\mathbb{F}_d$.
- To a composite physical system, having a finite number of possible states, we associate a state space which is the cartesian product of spaces of states of subsystems.
- Any physical process (state change) in a closed system is described by a permutation on the associated state space.
- Given a state of a physical system, the outcome of any observable is exactly specified and the measurement process does not affect the state.

1.2 The bit

The simplest physical system to consider is one with only two states, to which we associate $\mathbb{F}_2$ as state space. We refer to this system as a *bit*.

For n bits, we have the Hamming space $\mathbb{F}_2^n$ of cardinality 2^n. There, according to (1.1), the scalar product $u \cdot u$ defines the *parity* of u, i.e. whether u contains an even number of ones (parity 0) or an odd number of ones (parity 1). Correspondingly, $u \cdot v$ gives the parity of v with respect to u.

Proposition 1.2.1. *Given $u \in \mathbb{F}_2^n$, $u \neq 0$, the parity condition determined by it divides the Hamming space $\mathbb{F}_2^n$ into two halves,*

$$h_0 = \{v \in \mathbb{F}_2^n | u \cdot v = 0\}, \quad h_1 = \{v \in \mathbb{F}_2^n | u \cdot v = 1\},$$

such that $|h_0| = |h_1| = 2^{n-1}$.

Proof. The proof goes by induction over n. The proposition is trivially true for $n = 1$. For $n > 1$, $u \in \mathbb{F}_2^n$ can be written as $(u', 0)$ or $(u', 1)$, with $u' \in \mathbb{F}_2^{n-1}$. Now, let be $o' \in \mathbb{F}_2^{n-1}$ such that $u' \cdot o' = 1$ and $z' \in \mathbb{F}_2^{n-1}$ such that $u' \cdot z' = 0$.

On the one hand, if $u = (u', 0)$, we have

$$(u', 0) \cdot (z', 0) = (u', 0) \cdot (z', 1) = 0.$$

Since there are 2^{n-2} strings z', we will have 2^{n-1} strings giving zero parity check with $(u', 0)$ (obviously the other 2^{n-1} give parity check one).

On the other hand, if $u = (u', 1)$, we have

$$(u', 1) \cdot (o', 1) = (u', 1) \cdot (z', 0) = 0.$$

Since there are 2^{n-2} strings z' as well as 2^{n-2} strings o', we will have 2^{n-1} strings giving zero parity check with $(u', 1)$ (obviously the other 2^{n-1} give parity check one). □

In $\mathbb{F}_2^n$, the Hamming distance, Definition 1.1.1, results

$$d(u, v) = \mathrm{wt}(u \oplus v), \tag{1.2}$$

where wt is the so-called *weight* of a vector, the number of ones it contains.

Finally, we summarize the relations between bit and physical systems:

- $\mathbb{F}_2$ represents the state space of a physical system that can only have two possible states.
- The state space of n of such systems is $\mathbb{F}_2^n$.
- The only possible physical processes (state change) for a bit system are identity and bit-flip.
- Given the state 0 (resp., 1) of a bit, by measuring the observable we get 0 (resp., 1) with certainty, i.e. we are able to know the state of the system and the latter will be left unaffected by measurement.

Exercises

1. Prove the triangle inequality property for the Hamming distance, Definition 1.1.1.
2. Show that the parity of u can be equally written as $u \cdot \mathbf{1}$, where $\mathbf{1}$ is the string whose components are all equal to 1.
3. Prove the relation (1.2).

Chapter 2

INFORMATION AND ENTROPY: THE CLASSICAL VIEW

In this chapter, the central notion of *information* will be introduced by resorting, as is customary, to randomness. Information will then be characterized by means of several *entropies*. We are still confining the analysis to a classical approach.

2.1 Shannon entropy

The state of a physical system is not always known with certainty. It may happen that it is only known with some probability. In such a case, it is useful to resort to the notion of *random variable* X defined on an alphabet $\mathcal{X}$, i.e. a variable taking values $x \in \mathcal{X}$ with probability distribution (p.d.) $p_X(x) = \Pr\{X = x\}$, hence, $X \sim p_X(x)$. In the rest of the book, we often simply write $p(x)$ or p_x instead of $p_X(x)$.

Practically, we no longer describe the state of a system as an element of $\mathcal{X}$ (or $\mathbb{F}_d$) — a pure state — but rather by a probability vector $(p(x) : x \in \mathcal{X})$ — a mixed state. This is equivalent to dealing with an *ensemble of states* $\{x, p_x\}_{x \in \mathcal{X}}$.

Th *expectation value* of a real-valued function f of the random variable X, or more generally a function taking values in a real vector space, is defined as

$$\mathbb{E}(f(X)) := \sum_{x \in \mathcal{X}} p(x) f(x). \tag{2.1}$$

According to C. E. Shannon, the notion of *information* should be related to the uncertainty in knowing the state of a system (value of the random variable).

A measure of the information (gain) is thus a measure of (*a priori*) uncertainty. The latter in turn is given by the degree of "surprise". Shannon argued that the degree of surprise in getting the outcome x from the random variable X can be quantified by $-\log x$. Here and henceforth, log is intended in the base 2.

This measure of surprise ensures that it is large for unlikely outcomes and zero for certain outcomes. It also guarantees continuity and additivity for independent events. Then the average measure of surprise led Shannon to quantify information as follows:

Definition 2.1.1. The (Shannon) entropy of a discrete random variable X taking values $x \in \mathcal{X}$ with probability $p_X(x)$ is

$$H(X) := -\sum_{x \in \mathcal{X}} p(x) \log p(x) = \mathbb{E}(-\log p(x)).$$

In this definition, it is intended to put $0 \log 0 = 0$, according to the continuity argument $\lim_{\epsilon \to 0} \epsilon \log \epsilon = 0$. We can also simply indicate the Shannon entropy $H(X)$ as $H(\vec{p})$, where $\vec{p} = (p(x) : x \in \mathcal{X})$ is the associated probability vector.

Note that $H(X)$ is expressed in *bits*, by our choice of logarithm on base 2. Hence, in this context the *bit* becomes the basic unit to quantify information.

Actually, $H(X)$ quantifies how much information we gain (on average) when we learn the value of X. A prototypical example is provided by the binary entropy.

2.1.1 *The binary entropy*

Consider $X \in \mathcal{X} = \{0, 1\}$ and characterized by $\Pr\{X = 1\} = p$, $\Pr\{X = 0\} = 1 - p$. Then,

$$H(X) = -p \log p - (1 - p) \log(1 - p), \tag{2.2}$$

which is a function of only one parameter $p \in [0, 1]$, thus also indicated as $H(p)$ (or $H_2(p)$, to emphasize its binary character).

It has the following properties:

(1) it is symmetric, $H(p) = H(1-p)$;
(2) the minimum value is at $p = 0$ and $p = 1$; $H(0) = H(1) = 0$;
(3) the maximum value is at $p = \frac{1}{2}$; $H(\frac{1}{2}) = 1$;
(4) it is concave, i.e. it holds

$$H(qp_1 + (1-q)p_2) \geqslant qH(p_1) + (1-q)H(p_2), \tag{2.3}$$

for all p_1, p_2, $q \in [0,1]$.

2.1.2 *Jensen inequality*

The property (2.3) of the binary entropy is the concavity, a generally powerful concept.

Definition 2.1.2. A function $f : \mathbb{R}^n \to \mathbb{R}$ (or more generally $f : \mathcal{C} \to \mathbb{R}$, for a convex set $\mathcal{C} \subset \mathbb{R}^n$) is *concave* if

$$f(px_1 + (1-p)x_2) \geq pf(x_1) + (1-p)f(x_2),$$

for any $p \in [0,1]$. On the other hand, f is *convex* if $-f$ is concave, i.e.

$$f(px_1 + (1-p)x_2) \leq pf(x_1) + (1-p)f(x_2).$$

Example. $\log x$, $\sqrt{x}$ are concave functions on $\mathbb{R}_{\geq 0}$. $x \log x$, x^2 are convex functions on $\mathbb{R}_{\geq 0}$ and $\mathbb{R}$, respectively.

Proposition 2.1.3. *Let $f : \mathbb{R}^n \to \mathbb{R}$ be a concave function, then the following inequality holds, known as* Jensen's inequality

$$f\left(\sum_{i=1}^{n} p_i x_i\right) \geq \sum_{i=1}^{n} p_i f(x_i),$$

for any $p_i \geq 0$, with $\sum_{i=1}^{n} p_i = 1$. Clearly, the number of $p^i s$ can be understood as probabilities of a random variable X with $\Pr\{X = x_i\} = p_i$, *so that we can write the above equivalently*

$$f(\mathbb{E}(X)) \geq \mathbb{E}(f(X)).$$

Proof. By definition of concavity, the Jensen inequality holds true for $n = 2$, and trivially for $n = 1$. To prove it in general, let us assume that it is true for some $n \geq 2$ and we will show that it is also true for $n + 1$ (proof by induction). We can write

$$f\left(\sum_{i=1}^{n+1} p_i x_i\right) = f\left((1 - p_{n+1})\frac{\sum_{i=1}^{n} p_i x_i}{1 - p_{n+1}} + p_{n+1}x_{n+1}\right)$$

$$\geq (1 - p_{n+1})f\left(\frac{\sum_{i=1}^{n} p_i}{1 - p_{n+1}}x_i\right) + (1 - p_{n+1})f(x_{n+1})\,.$$

By noticing that the quantities $p_i' := \frac{p_i}{1-p_{n+1}}$ define a probability vector for $i = 1, \ldots, n$, and since we have assumed that the Jensen inequality holds true for n, we get

$$f\left(\sum_{i=1}^{n+1} p_i x_i\right) \geq (1 - p_{n+1})\sum_{i=1}^{n} \frac{p_i}{1 - p_{n+1}} f(x_i) + (1 - p_{n+1})f(x_{n+1})$$

$$= \sum_{i=1}^{n+1} p_i f(x_i)\,,$$

concluding the induction step. □

2.2 The relative entropy

Definition 2.2.1. Given two probability distributions p and q on the same alphabet $\mathcal{X}$, their relative entropy is defined as

$$D(p\|q) := \sum_{x\in\mathcal{X}} p(x)\log\frac{p(x)}{q(x)}.$$

Note that the relative entropy is well-defined only if the support of the probability p is contained in the support of the probability q. By convention, in all other cases we declare $D(p\|q) = +\infty$.

Although $D(\cdot\|\cdot)$ in many ways has the character of a distance between probability distributions, it is not a true distance since it is not symmetric and does not satisfy the triangle inequality. However,

Lemma 2.2.2. *$D(p\|q) \geqslant 0$, and equality holds if and only if $p = q$.*

Proof. First denote $\mathcal{X}_+ := \operatorname{supp} p = \{x \in \mathcal{X} : p(x) > 0\}$, then we can write

$$-D(p\|q) = -\sum_{x\in\mathcal{X}_+} p(x)\log\frac{p(x)}{q(x)} = \sum_{x\in\mathcal{X}_+} p(x)\log\frac{q(x)}{p(x)} \tag{2.4}$$

$$\leq \log\left(\sum_{x\in\mathcal{X}_+} p(x)\frac{q(x)}{p(x)}\right) = \log\left(\sum_{x\in\mathcal{X}_+} q(x)\right) \tag{2.5}$$

$$\leq \log\left(\sum_{x\in\mathcal{X}} q(x)\right) = \log 1 = 0. \tag{2.6}$$

where (2.5) is obtained from the Jensen's inequality (Proposition 2.1.3) by noticing that the logarithm is a concave function. Still in (2.5), the equality holds true if and only if $q(x)/p(x)$ is constant, i.e. $q(x) = \kappa p(x)$. Hence $\sum_{x\in\mathcal{X}_+} q(x) = \kappa\sum_{x\in\mathcal{X}_+} p(x) = \kappa$. On the other hand, in (2.6) we have the equality only if $\sum_{x\in\mathcal{X}_+} q(x) = \sum_{x\in\mathcal{X}} q(x) = 1$, which implies that $\kappa = 1$. Therefore we have $D(p\|q) = 0$ if and only if $p = q$. □

This also allows us to determine the range of the Shannon entropy $H(X)$. Let $u(x) := 1/|\mathcal{X}|$ be the uniform distribution, then

$$\begin{aligned} H(X) &= -D(p\|u) - \sum_x p(x)\log\left(1/|\mathcal{X}|\right) \\ &= -D(p\|u) + \log|\mathcal{X}| \\ &\leqslant \log|\mathcal{X}|, \end{aligned}$$

where we have used Lemma 2.2.2. The equality holds if and only if $p = u$. Hence,

$$0 \leq H(X) \leq \log|\mathcal{X}|, \tag{2.7}$$

where the minimum is reached when all the symbols have zero probability except one, and the maximum is reached for the uniform probability distribution.

2.3 Fidelity between probabilities

Knowing that relative entropy is not a proper distance (metric) between probabilities, let us look for another quantity.

Definition 2.3.1. The fidelity between two probability distributions p and q on $\mathcal{X}$ is defined as

$$F(p,q) := \sum_{x\in\mathcal{X}} \sqrt{p(x)q(x)}.$$

This is still not a metric because $F(p,p) = 1$. However, it can be interpreted as the scalar product on $\mathbb{R}^{\mathcal{X}}$ between vectors with components $\sqrt{p(x)}$ and $\sqrt{q(x)}$, which lie on a unit sphere. As such, it equals the cosine of the angle between vectors and this angle can be regarded as a distance.

Definition 2.3.2. The (angular) distance between two probability distributions p and q on $\mathcal{X}$ is defined as

$$d(p,q) := \arccos F(p,q).$$

2.4 Other entropic quantities

Let us consider a pair of random variables X, Y defined over alphabets $\mathcal{X}$, $\mathcal{Y}$ with joint probability

$$p(x,y) = \Pr\{X = x, Y = y\}.$$

Then, taking into account the conditional probabilities

$$p(x|y) = \Pr\{X = x|Y = y\},$$
$$p(y|x) = \Pr\{X = x|Y = y\},$$

the following relations (Bayes rules) hold true

$$p(x,y) = p(x|y)p(y),$$
$$p(x,y) = p(y|x)p(x),$$

where

$$p(x) = \sum_{y \in \mathcal{Y}} p(x, y),$$

$$p(y) = \sum_{x \in \mathcal{X}} p(x, y),$$

are the so-called *marginal* probability distributions.

The two random variables X and Y are (statistically) independent if and only if $p(x, y) = p(x)p(y)$ (otherwise they are correlated).

2.4.1 *Joint entropy*

Definition 2.4.1. Given two random variables X, Y defined over alphabets $\mathcal{X}$, $\mathcal{Y}$ with joint probability $p(x, y)$, we can consider the pair (X, Y) as a single random variable taking values in $\mathcal{X} \times \mathcal{Y}$, and their joint entropy is

$$H(X, Y) = -\sum_{x \in \mathcal{X}} \sum_{y \in \mathcal{Y}} p(x, y) \log p(x, y).$$

It quantifies the amount of (*a priori*) uncertainty about the two variables X, Y. From the properties (2.7) of the Shannon entropy, it follows that

$$0 \leq H(X, Y) \leq \log |\mathcal{X}| + \log |\mathcal{Y}|. \tag{2.8}$$

2.4.2 *Conditional entropy*

The conditional entropy $H(Y|X)$ quantifies the residual uncertainty about the variable Y once we know the value of the other variable X.

Definition 2.4.2. Given two random variables X, Y defined over alphabets $\mathcal{X}$, $\mathcal{Y}$, the conditional entropy of Y with respect to X is

$$H(X|Y) := H(X, Y) - H(Y).$$

Analogously, we can define $H(Y|X) = H(X,Y) - H(X)$. From the Bayes rules, we have $\log p(x,y) = \log p(y|x) + \log p(x)$, with the joint probability $p(x,y)$ and conditional probability $p(y|x)$. Thus,

$$-\sum_{x\in\mathcal{X}}\sum_{y\in\mathcal{Y}} p(x,y)\log p(x,y) = -\sum_{x\in\mathcal{X}}\sum_{y\in\mathcal{Y}} p(x,y)\log p(x|y) - \sum_{y\in\mathcal{X}} p(y)\log p(y),$$

that is

$$H(X|Y) = \sum_{y\in\mathcal{Y}} p(y)H(X|Y=y),$$

where $H(X|Y = y) = -\sum_{x\in\mathcal{X}} p(x|y)\log p(x|y)$ is the Shannon entropy of X conditional on Y taking the value $y \in \mathcal{Y}$.

Clearly, $H(X|Y) \geq 0$ because in the above equation $H(X|Y = y) \geq 0$, for all y.
Then, by using $H(Y|X) \geq 0$, we immediately arrive at

$$H(X,Y) \geq H(Y). \tag{2.9}$$

Analogously, we have $H(X,Y) \geq H(X)$.

2.4.3 *Mutual information*

The mutual information $I(X{:}Y)$ quantifies the amount of information that one random variable contains about another random variable, i.e. amount by which uncertainty of one random variable is reduced due to the knowledge of the other.

Definition 2.4.3. Given two random variables X, Y defined over alphabets $\mathcal{X}$, $\mathcal{Y}$, their mutual information is

$$I(X:Y) := H(X) + H(Y) - H(X,Y).$$

It immediately follows that

$$I(X:Y) = H(X) - H(X|Y)$$
$$= H(Y) - H(Y|X).$$

Furthermore, $I(X:Y) = I(Y:X)$ and $I(X:X) = H(X)$.

We can easily check that furthermore it holds

$$I(X:Y) = \sum_{x\in\mathcal{X}} \sum_{y\in\mathcal{Y}} p(x,y) \log \frac{p(x,y)}{p(x)p(y)},$$

or equivalently

$$I(X:Y) = D\left(p(x,y) \| p(x)q(y)\right),$$

i.e. the relative entropy between the joint probability $p(x,y)$ and the product of the marginals $p(x)p(y)$.

From the latter it follows that if X and Y are independent, then $I(X:Y) = 0$; otherwise, $I(X:Y) > 0$.

2.5 Further properties of entropies

We list here the most relevant properties of the above-introduced entropic quantities.

(1) *Subadditivity of the Shannon entropy*

$$H(X,Y) \leq H(X) + H(Y).$$

This follows from the fact that $I(X:Y) \geq 0$. If X and Y are independent, then $I(X:Y) = 0$ and hence $H(X,Y) = H(X) + H(Y)$, i.e. we have the additivity property.

(2) *Conditioning reduces entropy*

$$H(Y|X) \leq H(Y).$$

This follows from the fact that $I(X:Y) \geq 0$.

(3) *Concavity of the Shannon entropy*

$$H(\lambda p_X + (1-\lambda)p_{X'}) \geq \lambda H(p_X) + (1-\lambda)H(p_{X'}),$$

for $\lambda \in [0,1]$ and the distributions p_X, $p_{X'}$ of two random variables X, X' taking values in the same set $\mathcal{X}$. The space of probabilities on $\mathcal{X}$ (with $n = |\mathcal{X}|$) is the convex set

$$\mathfrak{P}(\mathcal{X}) = \left\{ \vec{p} = (p_1, \ldots, p_n) \middle| 0 \leq p_i \leq 1, \sum_{i=1}^{n} p_i = 1 \right\}.$$

To see the above, let $\vec{p} \in \mathfrak{P}$, and $\vec{q} = (q_1, \ldots, q_n) \in \mathbb{R}^n$ be such that $\sum_{i=1}^{n} q_i = 0$ and such that $\vec{p} + \epsilon\vec{q} \in \mathfrak{P}$ for ϵ is small enough. Then,

$$H(\vec{p} + \epsilon\vec{q}) = -\sum_{i=1}^{n} (p_i + \epsilon q_i) \log(p_i + \epsilon q_i)$$

$$\frac{dH(\vec{p} + \epsilon\vec{q})}{d\epsilon} = -\sum_{i=1}^{n} q_i \log(p_i + \epsilon q_i)$$

$$\frac{d^2H(\vec{p} + \epsilon\vec{q})}{d\epsilon^2} = -\frac{1}{\ln 2} \sum_{i=1}^{n} \frac{q_i^2}{p_i + \epsilon q_i} < 0,$$

because $\vec{p} + \epsilon\vec{q} \in \mathfrak{P}$.

(4) *Chain rule*

$$H(X,Y,Z) = H(X) + H(Y|X) + H(Z|X,Y).$$

This follows from writing

$$\begin{aligned} H(X,Y,Z) &= H(X,Y,Z) + H(X) - H(X) \\ &\quad + H(X,Y) - H(X,Y) \\ &= H(X) + H(Y|X) + H(Z|X,Y). \end{aligned}$$

(5) *Strong subadditivity*

$$H(X,Y,Z) + H(Y) \leq H(X,Y) + H(Y,Z).$$

This can be proved by extending the argument for property 1 to the conditional entropy, since it is equivalent to

$$H(X, Z|Y) \leq H(X|Y) + H(Z|Y).$$

Exercises

1. Given probability distributions p, q on $\mathcal{X} = \{0,1\}$, and the uniform distribution u, such that $F(u,p) > F(u,q)$, show that $H(p) > H(q)$. Is the same true for $|\mathcal{X}| > 2$?
2. Derive property 3 (concavity) of the Shannon entropy from property 2 (conditioning reduces the entropy), by defining a suitable pair of random variables.
3. Show that alternatively to Definition 2.3.2, also $d(p,q) := \sqrt{1 - F(p,q)^2}$ is a metric.
4. Let the random variables $X_1, X_2, \ldots, X_n$ be distributed according to $p(x_1, x_2, \ldots, x_n)$. Show that

$$H(X_1, X_2, \ldots, X_n) \leq \sum_{i=1}^{n} H(X_i),$$

with equality if and only if the X_i are independent.
5. Let X be a random variable on $\mathcal{X}$ and $f : \mathcal{X} \to \mathcal{Y}$ be a function. Show that the entropy of $Y = f(X)$ is less than or equal to the entropy of X, $H(Y) \leq H(X)$.
6. Show that for non-negative numbers $a_1, \ldots, a_n$ and $b_1, \ldots, b_n$, the log-concavity inequality holds

$$\sum_{i=1}^{n} a_i \log \frac{a_i}{b_i} \geq \left(\sum_{i=1}^{n} a_i \right) \log \frac{\sum_{i=1}^{n} a_i}{\sum_{i=1}^{n} b_i}.$$

Use this inequality to show that the relative entropy $H(p\|q)$ is convex in the pair (p,q), i.e.

$$H(\lambda p_1 + (1-\lambda)p_2 \| \lambda q_1 + (1-\lambda)q_2) \leq \lambda H(p_1\|q_1) + (1-\lambda)H(p_2\|q_2),$$

for all $0 \leq \lambda \leq 1$.

Chapter 3

A QUANTUM PRIMER

This chapter introduces the basic notions of quantum mechanics in a way to be immediately usable for information theoretical purposes, that is without dwelling on physical interpretations.

3.1 Postulates of quantum mechanics

Postulate 3.1.1. *Each isolated physical system has associated to it a complex Hilbert space $\mathcal{H}$. The states of the system are then represented by state vectors, i.e. unit vectors in the Hilbert space.*

Henceforth, we will only consider finite-dimensional Hilbert spaces $\mathcal{H} \simeq \mathbb{C}^n$. The (complex) dimension of $\mathcal{H}$ will be denoted by $|\mathcal{H}| = n$. Furthermore, the Dirac notation is commonly used to indicate vectors $|\psi\rangle$ (*ket*) and their dual $\langle\psi|$ (*bra*). We revisit the notions of linear algebra in the light of the Dirac formalism, in the Appendix.

Given two vectors $|\phi\rangle, |\psi\rangle \in \mathcal{H}$, their scalar product is written as $\langle\phi|\psi\rangle$ (*bra-ket* notation). State vectors are such that they have unit norm, i.e. $\langle\psi|\psi\rangle = 1$.

Postulate 3.1.2. *Each composite physical system has associated to it a Hilbert space that is the tensor product of the Hilbert spaces associated to its subsystems.*

Postulate 3.1.3. *The state vector change of an isolated physical system with associated Hilbert space $\mathcal{H}$ is described by a unitary* *operator* $U : \mathcal{H} \to \mathcal{H}$

$$|\psi\rangle \to |\psi'\rangle = U|\psi\rangle.$$

Postulate 3.1.4. *Each observable physical quantity* A *has associated to it to a Hermitian operator in the space of state vectors (Hilbert space* $\mathcal{H}$*).*

The possible measurement outcomes correspond to the eigenvalues $\{a_j\}_j$ $(a_j \in \mathbb{R})$ *of* A.

The probability that the outcome is a_j*, given that the system was in the state vector* $|\psi\rangle$*, is*

$$p_\psi(a_j) = \langle\psi|P_j|\psi\rangle,$$

with $P_j = |a_j\rangle\langle a_j|$ *the projector onto the subspace of eigenvector* $|a_j\rangle$ *of* A *corresponding to the eigenvalue* a_j.

As a consequence of getting the measurement outcome a_j*, the state vector of the system changes as*

$$|\psi\rangle \to |\psi'\rangle = \frac{P_j|\psi\rangle}{\langle\psi|P_j|\psi\rangle^{1/2}}.$$

Note that the expectation value of measuring A, when the state vector of the system is $|\psi\rangle$, results

$$\mathbb{E}(A) \equiv \langle A\rangle = \sum_j p_\psi(a_j)a_j = \langle\psi|A\psi\rangle.$$

It immediately follows that a uni-modular factor $e^{i\theta}$ $(\theta \in \mathbb{R})$ multiplying a state vector $|\psi\rangle$ will never be observed.

As a consequence of this and of the fact that we are focusing on unit norm vectors, we can identify proportional vectors, $|v\rangle \sim e^{i\theta}\frac{|v\rangle}{\||v\rangle\|}$ for any $|v\rangle \in \mathcal{H}$.

This is an equivalence relation that leads to considering the state of a system as given by a (projective) ray, i.e. an equivalence class, in $\mathcal{H}$, rather than as a vector. Analogously, the state space would be the projective space $\mathbb{P}(\mathcal{H})$, rather than the Hilbert space $\mathcal{H}$ itself. However, in the following, we will not use this specification and generically refer to Hilbert space as space of state vectors.

3.2 Quantum weirdness

3.2.1 *Entanglement*

Let us consider a composite system composed, for the sake of simplicity, of only two subsystems, i.e. a bipartite system with, according to Postulates 3.1.1 and 3.1.2, associated Hilbert space $\mathcal{H}_{AB} = \mathcal{H}_A \otimes \mathcal{H}_B$. Then we can distinguish two possible kinds of state vectors.

Definition 3.2.1. State vectors $|\Psi\rangle_{AB} \in \mathcal{H}_{AB}$ that can be written as $|\Psi\rangle_{AB} = |\psi\rangle_A \otimes |\phi\rangle_B$ with $|\psi\rangle_A \in \mathcal{H}_A$ and $|\psi\rangle_B \in \mathcal{H}_B$ are called *separable.*

Those state vectors $|\Psi\rangle_{AB} \in \mathcal{H}_{AB}$ that are not separable, i.e. $|\Psi\rangle_{AB} \neq |\psi\rangle_A \otimes |\phi\rangle_B$ for any $|\psi\rangle_A \in \mathcal{H}_A$ and $|\phi\rangle_B \in \mathcal{H}_B$, are called *entangled.*

Note that for separable state vectors, a well-defined state is assigned to each subsystem A and B. This is no longer true for entangled state vectors, for which the state vector is precisely assigned only to the global system AB.

Let us now consider a generic state vector $|\Psi\rangle_{AB} \in \mathcal{H}_{AB} = \mathcal{H}_A \otimes \mathcal{H}_B$. Let us put $|\mathcal{H}_A| = d_A$ and $|\mathcal{H}_B| = d_B$, and introduce the "local" bases $\{|i\rangle_A\}_{i=1,\ldots,d_A}$ and $\{|j\rangle_B\}_{j=1,\ldots,d_B}$ of $\mathcal{H}_A$ and $\mathcal{H}_B$, respectively. By expanding the "global" state vector in the basis $\{|ij\rangle = |i\rangle_A \otimes |j\rangle_B\}$, we get

$$|\Psi\rangle_{AB} = \sum_{i=1}^{d_A} \sum_{j=1}^{d_B} \Psi_{ij} |i\rangle_A |j\rangle_B,$$

with coefficients $\Psi_{ij} \in \mathbb{C}$ such that $\sum_{i,j} |\Psi_{ij}|^2 = 1$. These coefficients can be arranged in a matrix

$$\Psi \equiv \begin{pmatrix} \Psi_{11} & \Psi_{12} & \cdots & \Psi_{1d_B} \\ \Psi_{21} & \Psi_{22} & \cdots & \Psi_{2d_B} \\ \vdots & \vdots & \ddots & \vdots \\ \Psi_{d_A 1} & \Psi_{d_A 2} & \cdots & \Psi_{d_A d_B} \end{pmatrix}.$$

By the *singular-value decomposition* (see Appendix) we can write

$$\Psi = UDV,$$

where U is a $d_A \times d_A$ unitary matrix, V is a $d_B \times d_B$ unitary matrix, and D is a $d_A \times d_B$ matrix with non-negative entries on the principal diagonal and zero everywhere else. In matrix elements, the above equation reads

$$\Psi_{ij} = \sum_{a=1}^{\min\{d_A,d_B\}} U_{ia}\sqrt{\lambda_a}V_{aj},$$

where $\sqrt{\lambda_a}$ are the diagonal elements of D. Then we can rewrite the "global" state vector $|\Psi\rangle_{AB}$ as follows:

$$\begin{aligned}|\Psi\rangle_{AB} &= \sum_{i=1}^{d_A}\sum_{j=1}^{d_B}\sum_{a=1}^{\min\{d_A,d_B\}} U_{ia}\sqrt{\lambda_a}V_{aj}|i\rangle_A|j\rangle_B \\ &= \sum_{a=1}^{\min\{d_A,d_B\}} \sqrt{\lambda_a}\left(\sum_{i=1}^{d_A} U_{ia}|i\rangle_A\right)\left(\sum_{j=1}^{d_B} V_{aj}|j\rangle_B\right).\end{aligned}$$

Note that the state vectors $\sum_{i=1}^{d_A} U_{ia}|i\rangle_A =: |u_a\rangle_A$ and $\sum_{j=1}^{d_B} V_{bj}|j\rangle_B =: |v_b\rangle_B$ constitute elements of local bases in $\mathcal{H}_A$ and $\mathcal{H}_B$, respectively. Hence,

$$|\Psi\rangle_{AB} = \sum_{a=1}^{\min\{d_A,d_B\}} \sqrt{\lambda_a}|u_a\rangle_A|v_a\rangle_B. \tag{3.1}$$

The expression (3.1) is known as the *Schmidt decomposition* of a bipartite quantum state vector. The coefficients λ_a are called the *Schmidt coefficients* (they are unique up to reordering). Normalization implies that $\sum_a \lambda_a = 1$. Furthermore, the Schmidt coefficients turn out to be the eigenvalues of the matrix $\Psi^\dagger\Psi$.

The *Schmidt rank* is defined as the number of non-zero Schmidt coefficients (it equals the rank of the matrix Ψ).

The Schmidt decomposition gives a mathematical characterization of entangled state vectors:

(1) separable state vectors have Schmidt rank equal to one;
(2) entangled state vectors have Schmidt rank strictly larger than one.

The maximum Schmidt rank equals $d = \min\{d_A, d_B\}$.

Definition 3.2.2. State vectors $|\Psi\rangle_{AB} \in \mathcal{H}_{AB} = \mathcal{H}_A \otimes \mathcal{H}_B$ having maximum Schmidt rank and Schmidt coefficients all equal to $\frac{1}{d}$ are called *maximally entangled.*

3.2.2 *No-cloning*

In classical physics, given the state of a physical system, it is always possible to make a *copy* of it. To this end, consider the (permutation) map $\pi : \mathbb{F}_d \times \mathbb{F}_d \to \mathbb{F}_d \times \mathbb{F}_d$ such that

$$x, y \to \pi(x, y) = (x, (x \oplus y)); \quad x, y \in \mathbb{F}_d \times \mathbb{F}_d.$$

By simply taking $y = 0$, the above map gives a copy of x whatever is its value. Quantumly, it is not possible to devise a cloning map that works *universally* for any input state.

Definition 3.2.3. Given a Hilbert space $\mathcal{H}$, a universal cloning operator is a unitary operator $U : \mathcal{H} \otimes \mathcal{H} \to \mathcal{H} \otimes \mathcal{H}$ such that, for a fixed $|\alpha\rangle \in \mathcal{H}$, it realizes

$$U|\psi\rangle|\alpha\rangle = |\psi\rangle|\psi\rangle, \quad \forall|\psi\rangle \in \mathcal{H}.$$

Theorem 3.2.4. *A universal quantum cloning operator does not exist.*

Proof. Suppose that a universal cloning operator U exists. Then, considering $|\psi_1\rangle, |\psi_2\rangle \in \mathcal{H}$, we have

$$U|\psi_1\rangle|\alpha\rangle = |\psi_1\rangle|\psi_1\rangle,$$
$$U|\psi_2\rangle|\alpha\rangle = |\psi_2\rangle|\psi_2\rangle.$$

Taking the scalar product of the elements at left hand side (l.h.s.) and equating it to the scalar product of the elements at right hand side (r.h.s.), we get

$$\begin{aligned}\langle\alpha|\langle\psi_1|U^\dagger U|\psi_2\rangle|\alpha\rangle &= |\langle\psi_1|\psi_2\rangle|^2,\\ \langle\alpha|\alpha\rangle\langle\psi_1|\psi_2\rangle &= |\langle\psi_1|\psi_2\rangle|^2,\\ \langle\psi_1|\psi_2\rangle &= |\langle\psi_1|\psi_2\rangle|^2.\end{aligned}$$

The latter equality only occurs when $\langle\psi_1|\psi_2\rangle = 1$, i.e. the two vectors coincide, or when $\langle\psi_1|\psi_2\rangle = 0$, i.e. when the two vectors are orthogonal. □

3.2.3 *Distinguishing non-orthogonal states*

In classical physics, it is always possible (at least in principle) to measure a physical system without disturbing it (see Section 1.1). Hence, it would be always possible to perfectly distinguish between two states. In quantum mechanics, things are different.

Theorem 3.2.5. *Given $|\psi_1\rangle, |\psi_2\rangle \in \mathcal{H}$ such that $\langle\psi_1|\psi_2\rangle \neq 0, 1$, we cannot perfectly (i.e. with zero-error) distinguish between them with projective measurements (i.e. according to Postulate* 3.1.4).

Proof. Being $|\psi_1\rangle, |\psi_2\rangle$ non-orthogonal, they cannot be eigenvectors of an observable (self-adjoint operator) corresponding to distinct eigenvalues, so they cannot be perfectly distinguished. □

Corollary 3.2.6. *Any attempt to distinguish between $|\psi_1\rangle, |\psi_2\rangle \in \mathcal{H}$, such that $\langle\psi_1|\psi_2\rangle \neq 0, 1$, introduces a disturbance.*

Proof. Quite generally, we can imagine to add a second quantum system and perform a global unitary transformation on the two systems. This second system plays the role of a measurement apparatus. Let us consider its Hilbert space to be $\mathcal{H}_M$ and take a state vector $|u\rangle \in \mathcal{H}_M$. Then, given $U : \mathcal{H} \otimes \mathcal{H}_M \to \mathcal{H} \otimes \mathcal{H}_M$, we would like to have

$$\begin{aligned}|\psi_1\rangle|u\rangle \to U|\psi_1\rangle|u\rangle &= |\psi_1\rangle|v_1\rangle,\\ |\psi_2\rangle|u\rangle \to U|\psi_2\rangle|u\rangle &= |\psi_2\rangle|v_2\rangle,\end{aligned}$$

with $|v_1\rangle \neq |v_2\rangle$. In such a way, to learn the label 1 or 2 (though with non-zero-error), we could perform a measurement on the system $\mathcal{H}_M$ without perturbing the original system ($|\psi_i\rangle|v_i\rangle$ are factorable state vectors). However, proceeding like in the proof of Theorem 3.2.4, we arrive at

$$\langle\psi_1|\psi_2\rangle = \langle\psi_1|\psi_2\rangle\langle v_1|v_2\rangle.$$

The latter equation implies that either $|\psi_1\rangle$ and $|\psi_2\rangle$ are orthogonal, or $\langle v_1|v_2\rangle = 1$, that is, $|v_1\rangle = |v_2\rangle$. □

Actually, the case of orthogonal state vectors is the only one in which it is possible to perfectly distinguish them without disturbance. Then the following question arises: what would be a suitable measurement to perform in order to distinguish between two non-orthogonal state vectors, even in presence of error? We will address this issue in the subsequent chapter using more general tools.

3.3 The quantum bit

The simplest quantum system we can consider has a two-dimensional associated Hilbert space, i.e. $\mathcal{H} \simeq \mathbb{C}^2$. This is the quantum version of the classical bit with associated state space $\mathbb{F}_2$, and as such is called *quantum bit* (*qubit* for short).

In this space, one commonly defines the canonical basis as

$$|0\rangle := \begin{pmatrix} 1 \\ 0 \end{pmatrix}, \quad |1\rangle := \begin{pmatrix} 0 \\ 1 \end{pmatrix}. \tag{3.2}$$

Hence, a general qubit state is described by a vector

$$|\psi\rangle = \alpha|0\rangle + \beta|1\rangle, \tag{3.3}$$

with $\alpha, \beta \in \mathbb{C}$ and $|\alpha|^2 + |\beta|^2 = 1$. However, as the state is a (projective) ray in $\mathbb{C}^2$, we can parameterize α, β up to equivalence in such a way that

$$|\psi\rangle = \cos\left(\frac{\theta}{2}\right)|0\rangle + e^{i\varphi}\sin\left(\frac{\theta}{2}\right)|1\rangle, \tag{3.4}$$

where $\theta \in [0,\pi]$ and $\varphi \in [0,2\pi)$. This means that the set of states $\{|\psi(\theta,\varphi)\rangle\}_{\theta,\varphi}$ corresponds to the surface of a 2-sphere $\mathbb{S}^2 \subset \mathbb{R}^3$ of radius one (known as the Bloch sphere).

It is common to introduce the *Pauli operators* defined as

$$\begin{aligned}
&\sigma^X := |0\rangle\langle 1| + |1\rangle\langle 0|, \quad \sigma^Y := -i|0\rangle\langle 1| + i|1\rangle\langle 0|,\\
&\sigma^Z := |0\rangle\langle 0| - |1\rangle\langle 1|.
\end{aligned} \tag{3.5}$$

With their commutation relations

$$[\sigma^X,\sigma^Y] = 2i\sigma^Z, \quad [\sigma^X,\sigma^Z] = -2i\sigma^Y, \quad [\sigma^Y,\sigma^Z] = 2i\sigma^X,$$

they realize the $\mathfrak{su}(2)$ algebra. Moreover σ^X, σ^Y, σ^Z together with the identity operator I constitute a basis for $\mathfrak{L}(\mathbb{C}^2)$, the space of linear operators on $\mathbb{C}^2$.

The following is the matrix representation of Pauli operators in the canonical basis (3.2):

$$\sigma^X = \begin{pmatrix} 0 & 1 \\ 1 & 0 \end{pmatrix}, \quad \sigma^Y = \begin{pmatrix} 0 & -i \\ i & 0 \end{pmatrix}, \quad \sigma^Z = \begin{pmatrix} 1 & 0 \\ 0 & -1 \end{pmatrix}.$$

Note that $\{|0\rangle, |1\rangle\}$ are the eigenvectors of σ^Z with corresponding eigenvalues $\{+1,-1\}$. Analogously,

$$\left\{|+\rangle := \frac{1}{\sqrt{2}}(|0\rangle + |1\rangle), \quad |-\rangle := \frac{1}{\sqrt{2}}(|0\rangle - |1\rangle)\right\} \tag{3.6}$$

are the eigenvectors of σ^X with corresponding eigenvalues $\{+1,-1\}$.

The unitary operator that allows one to transform $\{|0\rangle, |1\rangle\}$ into $\{|+\rangle, |-\rangle\}$ is known as *Hadamard transform*. It has the following matrix representation in the canonical basis (3.2):

$$H = \frac{1}{\sqrt{2}}\begin{pmatrix} 1 & 1 \\ 1 & -1 \end{pmatrix}.$$

Now, by exponentiating the $\mathfrak{su}(2)$ algebra, we can get the special unitary group SU(2). Let us consider a linear combination of Pauli operators $\sigma_{\hat{n}} = n_x\sigma^X + n_y\sigma^Y + n_z\sigma^Z$, where $\hat{n} := (n_x, n_y, n_z) \in \mathbb{R}^3$ such that $n_x^2 + n_y^2 + n_z^2 = 1$. Then we define

$$\mathcal{R}_{\hat{n}}(\theta) := \exp\left(-i\frac{\theta}{2}\sigma_{\hat{n}}\right), \tag{3.7}$$

with $\theta \in [0, 2\pi]$. It results, by the Taylor expansion and the property $\sigma_{\hat{n}}^2 = I$, that $\mathcal{R}_{\hat{n}}(\theta) = \cos\left(\frac{\theta}{2}\right) I - i\sin\left(\frac{\theta}{2}\right)\sigma_{\hat{n}}$, and indeed $\{\mathcal{R}_{\hat{n}}(\theta)\}_{\hat{n},\theta} = \mathrm{SU}(2)$.

Hence, the operators $\mathcal{R}_{\hat{n}}(\theta)$ acting on vectors of $\mathbb{C}^2$ provide a representation (although not faithful) of rotations in $\mathbb{R}^3$. As a matter of fact, $\mathcal{R}_{\hat{n}}(\theta)$ rotates a generic vector $|\psi\rangle \in \mathbb{C}^2$ around $\hat{n}$ by an angle θ.

Example. $\mathcal{R}_{(0,1,0)}(\theta)$ takes $|\psi(0,0)\rangle \equiv |0\rangle$ into $|\psi(\theta,0)\rangle \equiv \cos\frac{\theta}{2}|0\rangle + \sin\frac{\theta}{2}|1\rangle$, thus rotating the initial vector (pointing along z direction) around y direction by an angle θ (if we assume θ, φ as spherical coordinates for $\mathbb{S}^2$).

As $\mathrm{U}(2) = \mathrm{U}(1) \times \mathrm{SU}(2)$, it turns out that any qubit unitary can be written, up to a unimodular factor, as $\mathcal{R}_{\hat{n}}(\theta)$ with suitable $\hat{n}$ and θ.

Since Pauli operators are Hermitian, according to Postulate 3.1.4, they also represent observable quantities.

Example. Suppose we measure the qubit observable σ^Z starting from the state vector (3.4). We will get outcomes $+1$ and -1 with the following probabilities and *a posteriori* state vectors:

$$\Pr\{+1\} = \cos^2\left(\frac{\theta}{2}\right), \quad |\psi'\rangle = |0\rangle,$$
$$\Pr\{-1\} = \sin^2\left(\frac{\theta}{2}\right), \quad |\psi'\rangle = |1\rangle.$$

Henceforth, we will use a simplified notation for outcomes of σ^Z measurements, namely

$$+1 \to 0, \quad -1 \to 1.$$

Then writing the $\mathbb{C}^2$ canonical basis as $\{|z\rangle\}_{z\in\mathbb{F}_2}$ and the vector state (3.4) as

$$|\psi\rangle = \sum_{z\in\mathbb{F}_2} c_z|z\rangle, \quad c_z \in \mathbb{C}, \quad \sum_{z\in\mathbb{F}_2} |c_z|^2 = 1,$$

we can say that the outcomes of σ^Z measurement are z with $\Pr\{z\} = |c_z|^2$. Following Postulate 3.1.2, this notation can be straightforwardly extended to a system composed by n qubits by simply considering $\mathbb{F}_2^n$ in place of $\mathbb{F}_2$.

For $n = 2$, besides the canonical basis $\{|z\rangle\}_{z\in\mathbb{F}_2^2}$, another interesting basis is given by the following state vectors:

$$\begin{aligned} |\Phi^\pm\rangle &:= \frac{1}{\sqrt{2}}\left(|00\rangle \pm |11\rangle\right), \\ |\Psi^\pm\rangle &:= \frac{1}{\sqrt{2}}\left(|01\rangle \pm |10\rangle\right), \end{aligned} \tag{3.8}$$

known as *Bell state vectors.* The unitary operator that allows one to transform the canonical basis into the Bell's basis is $CNOT(H \otimes I)$, where

$$CNOT := |0\rangle\langle 0| \otimes I + |1\rangle\langle 1| \otimes \sigma^X, \tag{3.9}$$

is the quantum version of the classical XOR logical gate.

3.4 Bits vs qubits

We compare here the main properties of bits and qubits.

- The state space for a bit is $\mathbb{F}_2 = \{0, 1\}$, or any two-element set.
  The state space for a qubit is $\mathbb{C}^2$ (with canonical basis state vectors $\{|z\rangle\}_{z\in\mathbb{F}_2}$ corresponding to bit states), or any two-dimensional Hilbert space.
- The state space for n bits is $\mathbb{F}_2^n$.
  The state space for n qubits is $\mathbb{C}^{2\otimes n}$ (with canonical basis state vectors $\{|z\rangle\}_{z\in\mathbb{F}_2^n}$ corresponding to n bits states). In case of entangled state vectors, we cannot assign a well-defined state vector to each qubit.

- Processes on a bit are described by permutations $\pi : \mathbb{F}_2 \to \mathbb{F}_2$.
  Processes on a qubit are described by unitaries $U : \mathbb{C}^2 \to \mathbb{C}^2$ (operators σ^X and I, a very small subclass of U(2), realize permutations on the canonical basis state vectors $\{|z\rangle\}_{z\in\mathbb{F}_2}$).
- Given a bit state $z \in \mathbb{F}_2$, the measurement of the only possible observable returns (with certainty) z and leaves the state unperturbed.
  Given a qubit state vector $\sum_{z\in\mathbb{F}_2} c_z|z\rangle$, the measurement of the observable σ^Z (one of infinitely many observables) returns $z \in \mathbb{F}_2$ with probability $\Pr\{z\} = |c_z|^2$, and projects the state vector onto $|z\rangle$. Thus, there is no way of reconstructing the qubit state vector by measurement of a single observable.

In conclusion, we may note that the qubit (Dirac) formalism encompasses the bit formalism, which can be recovered as a particular case.

Exercises

1. Let $f : \{0,1\}^2 \to \{0,1\}$ be a Boolean function, and define a state vector

$$|\psi_f\rangle := \frac{1}{2} \sum_{a,b\in\{0,1\}} (-1)^{f(a,b)} |a\rangle|b\rangle.$$

   Find the Schmidt number of $|\psi_f\rangle$ for $f = \text{AND}, \text{OR}, \text{XOR}$.
2. Given the two-qubit state vector

$$|\psi\rangle = \frac{1}{\sqrt{2}} \left(|01\rangle + |10\rangle\right),$$

   suppose we measure the first qubit along the $\hat{n}$-axis and the second qubit along the $\hat{m}$-axis, such that $\hat{n} \cdot \hat{m} = \cos\theta$. What is the probability of getting the pair $(+1, +1)$ as an outcome?
3. Given the Hadamard operator $H : \mathbb{C}^2 \to \mathbb{C}^2$ defined through the action on the canonical basis as $H|0\rangle = |+\rangle$ and $H|1\rangle = |-\rangle$, show

that its n-fold tensor product is represented by

$$H^{\otimes n} = \frac{1}{\sqrt{2^n}} \sum_{x,y\in\{0,1\}^n} (-1)^{x\cdot y} |x\rangle\langle y|$$

in the basis $\{|0\rangle, |1\rangle\}^{\otimes n}$, where $x \cdot y := x_1 y_1 \oplus x_2 y_2 \oplus \ldots \oplus x_n y_n$.

4. Using the state vectors

$$|\Phi_1\rangle = \frac{|00\rangle + |11\rangle}{\sqrt{2}}, \quad |\Phi_2\rangle = -i\frac{|00\rangle - |11\rangle}{\sqrt{2}},$$
$$|\Phi_3\rangle = \frac{|01\rangle - |10\rangle}{\sqrt{2}}, \quad |\Phi_4\rangle = -i\frac{|01\rangle + |10\rangle}{\sqrt{2}},$$

show that any two-qubit unitary can be written as

$$U = (V_1 \otimes V_2) \exp\left[-\frac{i}{2}(\alpha_x \sigma^X \otimes \sigma^X + \alpha_y \sigma^Y \otimes \sigma^Y + \alpha_z \sigma^Z \otimes \sigma^Z)\right](V_1' \otimes V_2'),$$

with qubit unitaries V_1, V_2, V_1', V_2'.

5. Generalize the Pauli operators σ^X and σ^Z to dimensions $d > 2$.

Chapter 4

MIXED QUANTUM STATES

This chapter may be considered as going over Chapter 3 and generalizing the concepts introduced there from pure to mixed states.

4.1 The density operator

In order to generalize the concept of an ensemble of classical states $\{x, p_x\}_{x\in\mathcal{X}}$, we may consider in the quantum framework an *ensemble of quantum state vectors* $\{|\psi_x\rangle, p_x\}_{x\in\mathcal{X}}$, with $|\psi_x\rangle$ belonging to a Hilbert space $\mathcal{H}$. Generally, the $|\psi_x\rangle$ do not need to be orthogonal, hence a faithful description of the ensemble cannot be provided by p_x. It is instead provided by the *density operator*, which plays a role analogous to the probability distribution in the classical setting.

Definition 4.1.1. Given a statistical mixture (ensemble) of quantum state vectors $\{p_i, |\psi_i\rangle\}_i$ on a Hilbert space $\mathcal{H}$, the corresponding density (statistical) operator is defined as $\rho := \sum_i p_i|\psi_i\rangle\langle\psi_i|$.

Theorem 4.1.2. *An operator* $\rho \in \mathfrak{L}(\mathcal{H})$ *is the density operator associated to some ensemble* $\{p_i, |\psi_i\rangle\}_i$ *if and only if* $\rho \geq 0$ *and* $\mathrm{Tr}(\rho) = 1$.

Proof. In the one direction, given that $\rho = \sum_i p_i|\psi_i\rangle\langle\psi_i|$, it is easy to show that $\mathrm{Tr}(\rho) = \sum_i p_i = 1$. Moreover, $\langle\varphi|\rho|\varphi\rangle = \sum_i p_i|\langle\varphi|\psi_i\rangle|^2 \geq 0$ for all $|\varphi\rangle \in \mathcal{H}$.

In the other direction, given that $\rho \geq 0$, it will also result Hermitian (the positive operators are a subclass of Hermitian operators). Hence, it possesses a spectral decomposition, say $\rho =$

$\sum_j \lambda_j |r_j\rangle\langle r_j|$, with orthonormal eigenvectors $\{|r_j\rangle\}_j$ and eigenvalues $\lambda_j \in \mathbb{R}_+$. Because $\mathrm{Tr}(\rho) = 1$, then $\sum_j \lambda_j = 1$. Thus, $\{\lambda_j, |r_j\rangle\}_j$ will be an ensemble of state vectors. □

It is worth remarking that different ensembles of quantum state vectors may give rise to the same density operator. In particular, this is the case for ensembles related by unitary transformations, as follows. Suppose that we can get $\{p_i, |\psi_i\rangle\}_i$ by applying a unitary U on another ensemble $\{q_j, |\varphi_j\rangle\}_j$, in the sense that

$$\sqrt{p_i}|\psi_i\rangle = \sum_j U_{ij}\sqrt{q_j}|\varphi_j\rangle,$$

then

$$\begin{aligned}
\rho &= \sum_i p_i|\psi_i\rangle\langle\psi_i| \\
&= \sum_i \sum_{jk} U_{ij}\sqrt{q_j}U^*_{ki}\sqrt{q_k}|\varphi_j\rangle\langle\varphi_k| \\
&= \sum_{jk} \delta_{jk}\sqrt{q_j q_k}|\varphi_j\rangle\langle\varphi_k| \\
&= \sum_j q_j|\varphi_j\rangle\langle\varphi_j|.
\end{aligned}$$

Definition 4.1.3. If the rank of the density operator $\rho \in \mathfrak{L}(\mathcal{H})$ is 1, the state of the system is said to be pure. If the rank of the density operator $\rho \in \mathfrak{L}(\mathcal{H})$ is strictly larger than 1, the state of the system is said to be mixed (maximally mixed if it coincides with $|\mathcal{H}|$ and all the eigenvalues are equal).

Note that the pure state corresponds to having a single state vector in the ensemble, i.e. $\rho = |\psi\rangle\langle\psi|$, which is unique up to global phase. On the contrary, a complete (or maximally) mixed state corresponds to having all the state vectors of an orthonormal basis with equal probability, i.e. $\rho = \frac{1}{d}I$.

It is easy to show that for pure states $\rho^2 = \rho$, hence $\mathrm{Tr}(\rho^2) = 1$, while for mixed states $\rho^2 \neq \rho$, hence $\mathrm{Tr}(\rho^2) < 1$.

The density operators on $\mathcal{H}$ form a convex subset (hereafter denoted as $\mathfrak{D}(\mathcal{H})$) of the vector space of $d \times d$ Hermitian matrices (which in turn is a subset of $\mathfrak{L}(\mathcal{H})$), and the pure states are precisely the extremal points of this subset.

4.2 Postulates of quantum mechanics revisited

The postulates of quantum mechanics formulated in Section 3.1 are now revisited in terms of density operators.

Postulate 4.2.1. *Each isolated physical system has associated to it a Hilbert space. The state of the system is then represented by a density operator on such Hilbert space.*

An arbitrary density operator of a qubit, i.e. on $\mathbb{C}^2$, can be written as

$$\rho = \frac{1}{2}(I + \sigma_{\vec{r}}), \tag{4.1}$$

where, similarly to Section 3.3, $\sigma_{\vec{r}} := r_x\sigma^X + r_y\sigma^Y + r_z\sigma^Z$ with $\vec{r} := (r_x, r_y, r_z) \in \mathbb{R}^3$, such that $r_x^2 + r_y^2 + r_z^2 \leq 1$. Hence, mixed states are inside the unit radius sphere $\mathbb{S}^2 \subset \mathbb{R}^3$, while pure states are on its surface.

Postulate 4.2.2. *Each composite physical system has associated to it a Hilbert space that is the tensor product of the Hilbert spaces associated to its subsystems.*

Postulate 4.2.3. *The state change of an isolated physical system with associated Hilbert space $\mathcal{H}$ is described by a unitary operator $U : \mathcal{H} \to \mathcal{H}$,*

$$\rho \to \rho' = U\rho U^\dagger.$$

Definition 4.2.4. Given a Hilbert space $\mathcal{H}$, a set of Hermitian operators $\{E_i\}_i \subset \mathfrak{L}(\mathcal{H})$, such that $E_i \geq 0$ and $\sum_i E_i = I$, forms a positive operator value measure (POVM).

Postulate 4.2.5. *A measurement process on $\mathcal{H}$ is described by means of a set of operators $\{M_m\}_m \subset \mathfrak{L}(\mathcal{H})$ giving rise to the POVM*

$\{M_m^\dagger M_m\}_m$. The possible measurement outcomes are labeled by the index "m" and occur with probability

$$\Pr\{m\} = \mathrm{Tr}(M_m^\dagger \rho M_m),$$

given that the prior state of the system was ρ.

As a consequence of the measurement outcome m, the state of the system changes as

$$\rho \to \rho' = \frac{M_m \rho M_m^\dagger}{\Pr\{m\}}.$$

Note that the projective measurements introduced in Postulate 3.1.4 are recovered for M_m, becoming orthogonal projectors, i.e. $M_m M_{m'} = \delta_{m,m'} M_m$. In such a case, they represent the projectors onto the eigenvectors subspaces of some Hermitian operator (observable) A whose expectation value reads $\mathbb{E}A \equiv \langle A \rangle = \mathrm{Tr}(A\rho)$.

4.3 Reduced density operator

Definition 4.3.1. Given a bipartite system AB with associated Hilbert space $\mathcal{H}_{AB} = \mathcal{H}_A \otimes \mathcal{H}_B$ in the state ρ_{AB}, the state of the subsystem A is given by $\rho_A = \mathrm{Tr}_B(\rho_{AB})$. The latter is referred to as reduced density operator of subsystem A.

Here, Tr_B stands for partial trace over B, i.e.

$$\mathrm{Tr}_B(\rho_{AB}) = \sum_{j=1}^{|\mathcal{H}_B|} {}_B\langle j|\rho_{AB}|j\rangle_B,$$

where $\{|j\rangle_B\}_j$ is an orthonormal basis of $\mathcal{H}_B$.

The notion of reduced density operator provides the correct measurement statistics for measurements done on a subsystem. As a matter of fact, consider an observable O_A for the subsystem A which we can write in $\mathcal{H}_{AB}$ as $O_A \otimes I_B$. Then

$$\begin{aligned}\mathrm{Tr}((O_A \otimes I_B)\rho_{AB}) &= \mathrm{Tr}\left((O_A \otimes I_B)\sum_{ij,i'j'} \rho_{ij,i'j'}|i\rangle_A\langle j| \otimes |i'\rangle_B\langle j'|\right) \\ &= \mathrm{Tr}\left(\sum_{ij,i'} \rho_{ij,i'} O_A |i\rangle_A\langle j|\right) \\ &= \mathrm{Tr}(O_A\rho_A).\end{aligned}$$

Now note that for maximally entangled states, which according to Definition 3.2.2 can be written as

$$|\psi\rangle_{AB} = \frac{1}{\sqrt{d}}\sum_{i=1}^{d} |i\rangle_A|i\rangle_B, \quad d = \min\{|\mathcal{H}_A|, |\mathcal{H}_B|\},$$

the reduced density operator results

$$\rho_A = \mathrm{Tr}_B(|\psi\rangle_{AB}\langle\psi|) = \frac{1}{d}I_A.$$

Similarly, we can find $\rho_B = \frac{1}{d}I_B$. In other words, the reduced density operator of a maximally entangled state is a maximally mixed state.

Given a mixed state of a system, it can always be considered as the reduced state of a pure state of a composite (larger) system.

Theorem 4.3.2 (Purification). *Given a mixed state ρ_A on $\mathcal{H}_A$, it is always possible to introduce a reference system with associated $\mathcal{H}_R \simeq \mathcal{H}_A$ and define a pure state $|\psi\rangle_{AR}\langle\psi|$, with $|\psi\rangle_{AR} \in \mathcal{H}_A \otimes \mathcal{H}_R$, such that $\rho_A = \mathrm{Tr}_R(|\psi\rangle_{AR}\langle\psi|)$.*

Proof. Suppose ρ_A has spectral decomposition $\rho_A = \sum_i p_i|i\rangle_A\langle i|$, then consider $\mathcal{H}_R = span\{|i\rangle_R\}_i \simeq \mathcal{H}_A$ and define $|\psi\rangle_{AR} := \sum_i \sqrt{p_i}|i\rangle_A|i\rangle_R$. It immediately follows that

$$\mathrm{Tr}_R(|\psi\rangle_{AR}\langle\psi|) = \sum_{ij} \sqrt{p_i p_j}|i\rangle_A\langle j|\mathrm{Tr}(|i\rangle_R\langle j|) = \rho_A. \qquad \square$$

4.4 More on quantum weirdness

4.4.1 *Entanglement in mixed states*

Definition 4.4.1. States ρ_{AB} on $\mathcal{H}_{AB} = \mathcal{H}_A \otimes \mathcal{H}_B$ that can be written as the product of states ρ_A on $\mathcal{H}_A$ and ρ_B on $\mathcal{H}_B$, i.e. $\rho_{AB} = \rho_A \otimes \rho_B$, are called *factorable.*

States ρ_{AB} on $\mathcal{H}_{AB} = \mathcal{H}_A \otimes \mathcal{H}_B$ that can be written as a convex combination of states ρ_A^k on $\mathcal{H}_A$ and ρ_B^k on $\mathcal{H}_B$, i.e. $\rho_{AB} = \sum_k p_k \rho_A^k \otimes \rho_B^k$, are called *separable.*

Those states ρ_{AB} on $\mathcal{H}_{AB} = \mathcal{H}_A \otimes \mathcal{H}_B$ that are not separable, i.e. $\rho_{AB} \neq \sum_k p_k \rho_A^k \otimes \rho_B^k$ for any states ρ_A^k on $\mathcal{H}_A$ and ρ_B^k on $\mathcal{H}_B$, are called *entangled.*

Note that this definition extends Definition 3.2.1.

In contrast to the case of pure states, it is in general not easy to check if a mixed state is separable or entangled. However, we have the following necessary criterion.

Theorem 4.4.2. *Given a density operator ρ_{AB} on $\mathcal{H}_{AB} = \mathcal{H}_A \otimes \mathcal{H}_B$ a necessary condition for its separability is that $\rho^{\top_A} \geq 0$, where $\top_A$ denotes the transposition with respect only to the subsystem A (so-called partial transposition)*[1].

Proof. By hypothesis ρ_{AB} is separable, hence it can be written as

$$\rho_{AB} = \sum_k p_k \rho_A^k \otimes \rho_B^k,$$

thus we have

$$\rho_{AB}^{\top_A} = \sum_k p_k (\rho_A^k)^\top \otimes \rho_B^k,$$

where $(\rho_A^k)^\top$ is the transpose of ρ_A^k. The latter being positive, also its transpose $(\rho_A^k)^\top$ will be positive. □

[1]Given a matrix representation of density operator ρ_{AB} with elements $\rho^{AB}_{i,i';j,j'} = {}_A\langle i|_B\langle j|\rho^{AB}|j'\rangle_B|i'\rangle_A$, it holds $(\rho^{AB}_{i,i';jj'})^\top = \rho^{AB}_{i',i;jj'}$.

It is possible to prove that the criterion of *positive partial transpose* is necessary *and* sufficient for a system of two qubits (Hilbert space isomorphic to $\mathbb{C}^2 \otimes \mathbb{C}^2$).

Example. Consider the Bell (hence, maximally entangled) state vector $|\Phi^+\rangle_{AB} = (|00\rangle + |11\rangle)/2$ giving rise to $\rho_{AB} = \frac{1}{2}|00\rangle\langle 00| + \frac{1}{2}|00\rangle\langle 11| + \frac{1}{2}|11\rangle\langle 00| + \frac{1}{2}|11\rangle\langle 11|$. It will result $\rho_{AB}^{\top_A} = \frac{1}{2}|00\rangle\langle 00| + \frac{1}{2}|01\rangle\langle 10| + \frac{1}{2}|10\rangle\langle 01| + \frac{1}{2}|11\rangle\langle 11|$. While the eigenvalues ρ_{AB} are $\{\frac{1}{2}, 0, 0, 0\}$, those of $\rho_{AB}^{\top_A}$ turn out to be $\{-\frac{1}{2}, \frac{1}{2}, \frac{1}{2}, \frac{1}{2}\}$.

4.4.2 *Distinguishing non-orthogonal states by POVMs*

We now generalize Theorem 3.2.5 by using POVMs.

Theorem 4.4.3. *Given $|\psi_1\rangle, |\psi_2\rangle \in \mathcal{H}$ such that $\langle\psi_1|\psi_2\rangle \neq 0, 1$ we cannot perfectly distinguish them without disturbance even by using POVM.*

Proof. Let us suppose that there exists a POVM with elements E_1, E_2 allowing to perfectly distinguish between two non-orthogonal state vectors $|\psi_1\rangle$, $|\psi_2\rangle$. We then have to require that

$$\langle\psi_1|E_1|\psi_1\rangle = 1, \quad \langle\psi_2|E_2|\psi_2\rangle = 1. \tag{4.2}$$

Since $E_1 + E_2 = I$, we have that $\langle\psi_1|E_2|\psi_1\rangle = 0$, which implies $\sqrt{E}_2|\psi_1\rangle = 0$. Let us decompose

$$|\psi_2\rangle = \alpha|\psi_1\rangle + \beta|\psi_1^\perp\rangle,$$

where $|\psi_1^\perp\rangle$ is orthogonal to $|\psi_1\rangle$ and $|\alpha|^2 + |\beta|^2 = 1$ with $|\beta|^2 < 1$. It follows that

$$\sqrt{E}_2|\psi_2\rangle = \beta\sqrt{E_2}|\psi_1^\perp\rangle$$

which implies

$$\langle\psi_2|E_2|\psi_2\rangle = |\beta|^2\langle\psi_1^\perp|E_2|\psi_1^\perp\rangle < 1,$$

which is in contradiction with (4.2). □

In conclusion, non-orthogonal state vectors cannot be perfectly distinguished, even allowing the most general form of measurement represented by POVM. We then may ask what is the best we can do to this end. In order to answer this question, we also have to consider the probability of appearance of the two non-orthogonal state vectors.

Suppose we have $|\psi_1\rangle$ and $|\psi_2\rangle$ (with $|\psi_1\rangle, |\psi_2\rangle \in \mathcal{H}$) each with probability 1/2 and furthermore the angle between them is $\theta \in [0, \pi/2]$, i.e. $|\langle\psi_1|\psi_2\rangle| = \cos\theta$.

If we want to avoid errors (unambiguous discrimination), the best we can do is to guess $|\psi_1\rangle$ (resp. $|\psi_2\rangle$) when projecting onto the subspace orthogonal to $|\psi_2\rangle$ (resp. $|\psi_1\rangle$). This however implies also the existence of an inconclusive answer (like "don't know"). The situation is described by the following POVM

$$E_1 = \frac{I - |\psi_2\rangle\langle\psi_2|}{1 + |\langle\psi_1|\psi_2\rangle|}, \quad E_2 = \frac{I - |\psi_1\rangle\langle\psi_1|}{1 + |\langle\psi_1|\psi_2\rangle|}, \quad E_0 = I - E_1 - E_2.$$

Note that E_1 and E_2 are proportional to projectors and the proportionality constant is necessary to guarantee that E_0 is also semidefinite positive.

The probability of success can be computed by considering the *a priori* state as

$$\rho = \frac{1}{2}|\psi_1\rangle\langle\psi_1| + \frac{1}{2}|\psi_2\rangle\langle\psi_2|, \tag{4.3}$$

and then

$$p(1) = \mathrm{Tr}(E_1\rho) = \frac{1}{2}(1 - \cos\theta), \tag{4.4}$$

$$p(2) = \mathrm{Tr}(E_2\rho) = \frac{1}{2}(1 - \cos\theta), \tag{4.5}$$

$$p(0) = \mathrm{Tr}(E_0\rho) = \cos\theta. \tag{4.6}$$

It results $P_{\text{succ}} = 1 - \cos\theta$.

Now, the average information gain is given by the difference between the initial uncertainty and the final uncertainty averaged overall possible outcomes. Actually, the initial uncertainty about $|\psi_1\rangle$ and $|\psi_2\rangle$ occurring with the same probability 1/2 is, according to the binary Shannon entropy, equal to 1. For what concerns the final uncertainty, with this strategy it only relies on the POVM element

E_0. After such an inconclusive answer, the uncertainty is still 1, but this happens with probability $p(0) = \cos\theta$. Hence, the average information gain of this strategy is

$$\Delta H_{\text{ave}} = 1 - \cos\theta. \tag{4.7}$$

A larger gain can be achieved if we are willing to accept probabilities of errors, rather than occasional certainties mixed with inconclusive answers. To this end, we consider measurement of the following observable:

$$|\psi_1\rangle\langle\psi_1| - |\psi_2\rangle\langle\psi_2|. \tag{4.8}$$

In the basis $\{|\psi_1\rangle, |\psi_1^\perp\rangle\}$, it can be represented by the matrix

$$\begin{pmatrix} 1 & 0 \\ 0 & 0 \end{pmatrix} - \begin{pmatrix} \sin^2\theta & \sin\theta\cos\theta \\ \sin\theta\cos\theta & \sin^2\theta \end{pmatrix}, \tag{4.9}$$

whose (normalized) eigenvectors are

$$|v_+\rangle = -\frac{1+\sin\theta}{\sqrt{2+2\sin\theta}}|\psi_1\rangle + \frac{\cos\theta}{\sqrt{2+2\sin\theta}}|\psi_1^\perp\rangle, \tag{4.10}$$

$$|v_-\rangle = \frac{1-\sin\theta}{\sqrt{2-2\sin\theta}}|\psi_1\rangle + \frac{\cos\theta}{\sqrt{2-2\sin\theta}}|\psi_1^\perp\rangle, \tag{4.11}$$

corresponding to eigenvalues $\pm\sin\theta$. Then, with outcome $+\sin\theta$, we guess $|\psi_1\rangle$, while with outcome $-\sin\theta$, we guess $|\psi_2\rangle$. The probability of success reads

$$P_{\text{succ}} = \frac{1}{2}|\langle\psi_1|v_+\rangle|^2 + \frac{1}{2}|\langle\psi_2|v_-\rangle|^2 = \frac{1}{2}(1+\sin\theta). \tag{4.12}$$

With this strategy, the final uncertainty only relies on the probability vector $(P_{\text{succ}}, 1 - P_{\text{succ}})$. Thus, the average information gain reads

$$\begin{aligned} \Delta H_{\text{ave}} &= 1 - H(P_{\text{succ}}, 1 - P_{\text{succ}}) \\ &= \frac{1}{2}[(1+\sin\theta)\log(1+\sin\theta) + (1-\sin\theta)\log(1-\sin\theta)]. \end{aligned} \tag{4.13}$$

It is worth remarking that the quantity in (4.13) is always greater than the quantity in (4.7).

In the following, we generalize the results to the case of mixed states discrimination.

Assume to have ρ_1 or ρ_2 each with probability 1/2. Our goal is to come up with a strategy to maximize the success probability.

Suppose initially that ρ_1 and ρ_2 commute and have spectral representations $\rho_1 = \sum_i p_i |\varphi_i\rangle\langle\varphi_i|$ and $\rho_2 = \sum_i q_i |\varphi_i\rangle\langle\varphi_i|$. The optimal strategy is to measure in the basis $\{|\varphi_i\rangle\}_i$, and the problem of discriminating ρ_1 and ρ_2 reduces to a purely classical problem. Namely, let $A := \{i | p_i \geq q_i\}$. It is optimal to guess ρ_1 if and only if the outcome is in A. The success probability turns out to be

$$\begin{aligned} P_{\text{succ}} &= \frac{1}{2}\Pr[i \in A | \rho = \rho_1] + \frac{1}{2}\Pr[i \notin A | \rho = \rho_2] \\ &= \frac{1}{2}\sum_{i \in A} p_i + \frac{1}{2}\sum_{i \notin A} q_i \\ &= \frac{1}{2}\sum_i \max\{p_i, q_i\} \\ &= \frac{1}{2} + \frac{1}{2}\text{TV}(\{p_i\}, \{q_i\}). \end{aligned} \tag{4.14}$$

The latter quantity is the statistical (also known as *total variation*) distance between probability distributions $\{p_i\}$ and $\{q_i\}$, defined by $\text{TV}(\{p_i\}, \{q_i\}) := \frac{1}{2}\sum_i |p_i - q_i|$.

Now for the general case (non-commuting density operators), we have the following theorem.

Theorem 4.4.4. *The best success probability to discriminate two mixed states represented by ρ_1 or ρ_2 is given by*

$$P_{\text{succ}} = \frac{1}{2} + \frac{1}{2}\left(\frac{1}{2}\|\rho_1 - \rho_2\|_1\right).$$

Here, $\|\bullet\|_1$ is the Schatten 1-norm, also called trace norm (see Appendix). For $A \in \mathfrak{L}(\mathcal{H})$, it is defined as

$$\|A\|_1 := \text{Tr}\sqrt{A^\dagger A}. \tag{4.15}$$

It results $\|A\|_1 = \sum_i |\lambda_i|$ the λ_i are the singular values of A.

Proof. Consider a POVM $\{E_1, E_2\}$. Our strategy is to guess ρ_i when the outcome is 'i'. The success probability can be expressed as

$$\begin{aligned}
&\frac{1}{2}\mathrm{Tr}(E_1\rho_1) + \frac{1}{2}\mathrm{Tr}(E_2\rho_2)\\
&= \frac{1}{4}\mathrm{Tr}((E_1 + E_2)(\rho_1 + \rho_2)) + \frac{1}{4}\mathrm{Tr}((E_1 - E_2)(\rho_1 - \rho_2))\\
&=: \frac{1}{2}T_1 + \frac{1}{2}T_2. \qquad (4.16)
\end{aligned}$$

Now note that as $E_1 + E_2 = I$, $T_1 = 1$. Furthermore, by the Holder inequality for operators (see Appendix), we have the following:

$$T_2 = \frac{1}{2}\mathrm{Tr}((E_1 - E_2)(\rho_1 - \rho_2)) \leq \frac{1}{2}\|E_1 - E_2\|_\infty \|\rho_1 - \rho_2\|_1. \quad (4.17)$$

Since $0 \leq E_1, E_2 \leq I$, $\|E_1 - E_2\|_\infty \leq 1$, hence $T_1 + T_2 \leq 1 + \frac{1}{2}\|\rho_1 - \rho_2\|_1$ gives the upper bound to the success probability.

To prove that the bound is attainable, suppose $\Delta := \rho_1 - \rho_2$ has some eigenvalues $\lambda_1, \ldots, \lambda_m \geq 0$ and $\mu_1, \ldots, \mu_{m'} < 0$. Let $\mathcal{H}_+$ be the eigenspace associated to $\lambda_1, \ldots, \lambda_m$ and $\mathcal{H}_-$ be the eigenspace associated to $\mu_1, \ldots, \mu_{m'}$. Let E_1, E_2 be the projection into $\mathcal{H}_+$ and $\mathcal{H}_-$, respectively. Then $E_1\Delta$ has eigenvalues $\lambda_1, \ldots, \lambda_m$ and $E_2\Delta$ has eigenvalues $\mu_1, \ldots, \mu_{m'}$. We check that $T_2 \leq \frac{1}{2}\|\Delta\|_1$ can achieve the equality:

$$\begin{aligned}
T_2 &= \frac{1}{2}\mathrm{Tr}((E_1 - E_2)\Delta) = \frac{1}{2}\mathrm{Tr}(E_1\Delta) - \frac{1}{2}\mathrm{Tr}(E_2\Delta)\\
&= \frac{1}{2}\sum_i \lambda_i - \frac{1}{2}\sum_i \mu_i = \frac{1}{2}\|\Delta\|_1. \qquad (4.18)
\end{aligned}$$

□

It is easy to recover the previous result about pure states $|\psi_1\rangle$ and $|\psi_2\rangle$ by applying Theorem 4.4.4. In fact, according to the eigenvalues of (4.9), we have $\frac{1}{2}\|\rho_1 - \rho_2\|_1 = \sin\theta$.

Exercises

1. Given the following two-qubit state vector:

$$|\Psi\rangle = \sqrt{\frac{1}{3}}|+\rangle_A|+\rangle_B + \sqrt{\frac{2}{3}}|-\rangle_A|-\rangle_B,$$

consider the measurement of the observable σ_A^Z. Write down the *a-posteriori* state conditioned to each outcome and the corresponding probability. Also write the unconditional state.

2. In $\mathbb{C}^n$, given an orthonormal basis $\{|0\rangle, |1\rangle, \ldots, |n-1\rangle\}$, consider the following state vectors:

$$|\beta\rangle = \left(1 - \sum_{k=0}^{n-1} x_k\right)^{1/2} |0\rangle + \sum_{k=1}^{n-1} \sqrt{x_k} e^{i\phi_k} |k\rangle,$$

where $\phi_k \in [0, 2\pi)$ and $0 \leq x_k \leq 1$ with the constraints $0 \leq x_j \leq 1 - \sum_{k=j+1}^{n-1} x_k$. Compute $\rho(\beta) = |\beta\rangle\langle\beta|$ for $n = 4$ and show that the resolution of the identity $\int_\Omega d\mu(\beta)\rho(\beta) = I_4$ is satisfied: Here, $d\mu(\beta)$ is the measure

$$\frac{n!}{(2\pi)^{n-1}} \prod_{k=0}^{n-1} dx_k d\phi_k,$$

and Ω is the domain for ϕ_k and x_k. Finally, still considering $n = 4$, find the reduced density matrix for one qubit and a condition (on ϕ_k and x_k) for entanglement of the two qubits (identifying the ket labels as $0 \equiv 00$, $1 \equiv 01$, $2 \equiv 10$, $3 \equiv 11$).

3. Prove the following entropic uncertainty relation for two observables A and B on a Hilbert space $\mathcal{H}$:

$$H(A) + H(B) \geq -2 \log \frac{1 + \bar{f}}{2}, \quad \bar{f} = \max_{a,b} |\langle a|b\rangle|,$$

where $H(A)$ (resp., $H(B)$) stands for the Shannon entropy of the probability distribution associated to the measurement of A (resp., B) having eigenvectors $\{|a\rangle\}$ (resp., $\{|b\rangle\}$) and eigenvalues $\{a\}$ (resp., $\{b\}$).

4. Consider a qubit POVM whose elements are

$$M_1 = \frac{1}{2}|0\rangle\langle 0|, \quad M_2 = \frac{1}{2}|1\rangle\langle 1|,$$
$$M_3 = \frac{1}{2}|+\rangle\langle +|, \quad M_4 = \frac{1}{2}|-\rangle\langle -|.$$

Show how this measurement can be realized as projective measurement in a two-qubit Hilbert space by considering the second qubit as ancillary (this is a special case of the so-called Naimark theorem).

5. Consider a source emitting state vectors $|\psi_0\rangle, |\psi_1\rangle \in \mathcal{H}_d$ with probability p_0, p_1, respectively. Determine the minimum error probability for distinguishing the two state vectors with a binary POVM E_0, E_1.

Chapter 5

INFORMATION AND ENTROPY: THE QUANTUM VIEW

This chapter provides a quantum leap from the notions introduced in Chapter 2. Specifically, quantum information will be related to the uncertainty in knowing the state of a system, and to characterize it, quantum entropies will be introduced.

5.1 Von Neumann entropy

A random variable X, taking values on an alphabet $\mathcal{X}$ with probability $p(x)$ for $x \in \mathcal{X}$, can be viewed as an *ensemble* of (classical) states $\{x, p(x)\}$.

Let us now consider the Hilbert space $\mathcal{H} = span\{|x\rangle\}_{x\in\mathcal{X}}$ with the basis state vectors $|x\rangle$ representing classical states. Then, the ensemble of states (or equivalently the random variable X) can be written as $\{|x\rangle, p(x)\}$.

More generally, we could take an ensemble of state vectors $\{|\psi_x\rangle, p(x)\}$ with $|\psi_x\rangle \in \mathcal{H}$ not necessarily orthogonal. The mixed state $\rho = \sum_x p(x)|\psi_x\rangle\langle\psi_x|$ will be associated to the ensemble. Thus, the density operator ρ on $\mathcal{H}$ plays the role of the distribution of a classical random variable X. This raises the question of how to quantify the information content of ρ; the proper notion is the von Neumann entropy.

Definition 5.1.1. For a density operator $\rho \in \mathfrak{D}(\mathcal{H})$, the von Neumann entropy is defined as

$$S(\rho) := -\mathrm{Tr}(\rho \log \rho).$$

Suppose that $\{|\varphi_i\rangle\}_{i=1}^d$ is an orthonormal basis diagonalizing ρ, so that we have the spectral decomposition

$$\rho = \sum_{i=1}^{d} \lambda_i |\varphi_j\rangle\langle\varphi_i|,$$

with $\lambda_i \geq 0$ and $\sum_i \lambda_i = 1$. Then,

$$S(\rho) = -\sum_i \lambda_i \log \lambda_i \equiv H(\vec{\lambda}) \tag{5.1}$$

can be understood as the Shannon entropy of its eigenvalues.

A first trivial property of the von Neumann entropy is that $S(\rho) \geq 0$ with equality holding if and only if ρ is pure.

Furthermore, $S(\rho)$ is invariant under unitary operations, $S(U\rho U^\dagger) = S(\rho)$. This is obvious because the spectrum of an operator remains invariant under unitary transformations.

5.2 Quantum relative entropy

Definition 5.2.1. For any two density operators ρ, $\sigma \in \mathfrak{D}(\mathcal{H})$, the quantum relative entropy is defined as

$$D(\rho\|\sigma) := \mathrm{Tr}(\rho(\log\rho - \log\sigma)).$$

Note that $D(\rho\|\sigma)$ is well-defined only if *supp* $\rho \subseteq$ *supp* σ. In all other cases, we conventionally set $D(\rho\|\sigma) = +\infty$.

Theorem 5.2.2 (Klein's inequality). *The relative entropy is non-negative, i.e.* $D(\rho\|\sigma) \geqslant 0$.

Proof. Let $\rho = \sum p_i |\varphi_i\rangle\langle\varphi_i|$, $\sigma = \sum q_j |\xi_j\rangle\langle\xi_j|$ be the spectral decompositions of ρ and σ, respectively. Then

$$D(\rho\|\sigma) = \sum_i p_i \log p_i - \sum_i \langle\varphi_i|\rho\log\sigma|\varphi_i\rangle. \tag{5.2}$$

Now,

$$\begin{aligned}\langle\varphi_i|\rho\log\sigma|\varphi_i\rangle &= p_i\langle\varphi_i|\log\sigma|\varphi_i\rangle\\ &= \sum_j p_i\langle\varphi_i|\xi_j\rangle\langle\xi_j|\varphi_i\rangle\log q_j\\ &= \sum_j p_i P_{ij}\log q_j,\end{aligned}$$

with $P_{ij} \geq 0$ and $\sum_i P_{ij} = \sum_j P_{ij} = 1$. It follows from (5.2) that

$$D(\rho\|\sigma) = \sum_i p_i \log p_i - \sum_{ij} p_i P_{ij} \log q_j. \tag{5.3}$$

Recall that the logarithm is a concave function, i.e.

$$\sum_j P_{ij}\log q_j \leqslant \log\left[\sum_j P_{ij}q_j\right]. \tag{5.4}$$

Then, defining $r_i := \sum_j P_{ij}q_j$, we have from (5.4)

$$D(\rho\|\sigma) \geqslant \sum_i p_i \log_2 p_i - \sum_i p_i \log_2 r_i \equiv D(p\|r) \geqslant 0. \qquad \square$$

Corollary 5.2.3. *If $\rho \in \mathfrak{D}(\mathcal{H})$ with $|\mathcal{H}| = d$, then $S(\rho) \leqslant \log d$.*

Proof. Let us take in Theorem 5.2.2 $\sigma = \frac{1}{d}I$, then

$$D(\rho\|\sigma) = D(\rho\|\frac{1}{d}I) = \mathrm{Tr}\left[\rho\left(\log\rho - \log\frac{1}{d}I\right)\right] = -S(\rho) + \log d,$$

hence, by Klein's inequality, it follows that $S(\rho) \leq \log d$ with equality holding if and only if $\rho = \frac{1}{d}I$. $\square$

5.3 Fidelity between quantum states

Like the classical relative entropy, the quantum relative entropy is not a true distance measure in the space of density operators. Let us thus introduce the fidelity.

Definition 5.3.1. Given two density operators ρ, $\sigma \in \mathfrak{D}(\mathcal{H})$, their fidelity is defined as

$$F(\rho,\sigma) := \mathrm{Tr}\sqrt{\rho^{1/2}\sigma\rho^{1/2}}.$$

It is worth remarking that in case ρ and σ commute, they can be diagonalized in the same basis, say $\{|i\rangle\}_i$,

$$\rho = \sum_i r_i |i\rangle\langle i|, \quad \sigma = \sum_i s_i |i\rangle\langle i|,$$

and hence the fidelity results

$$F(\rho,\sigma) = F(r,s).$$

Furthermore, it holds

$$F(|\psi\rangle\langle\psi|,\sigma) = \sqrt{\langle\psi|\sigma|\psi\rangle},$$

from which it follows that

$$F(|\psi\rangle\langle\psi|,|\phi\rangle\langle\phi|) \equiv F(|\psi\rangle,|\phi\rangle) = |\langle\psi|\phi\rangle|.$$

The definition of fidelity 5.3.1 is motivated by the following theorem (given without proof).

Theorem 5.3.2 (Uhlmann's theorem). *Given density operators ρ, $\sigma \in \mathfrak{D}(\mathcal{H}_Q)$, let us introduce an auxiliary (reference) system R such that $\mathcal{H}_R \simeq \mathcal{H}_Q$, then*

$$F(\rho,\sigma) = \max_{|\psi\rangle,|\phi\rangle} |\langle\psi|\phi\rangle|,$$

where $|\psi\rangle$ and $|\phi\rangle$ run over purifications of ρ and σ in $\mathcal{H}_Q \otimes \mathcal{H}_R$, respectively.

It is then clear that the fidelity is symmetric, however like the fidelity between probabilities, it is not a distance because $F(\rho,\rho) = 1$. Nevertheless, we can define the (angular) distance (also known as Bures angle) between density operators as

$$d(\rho,\sigma) := \arccos F(\rho,\sigma). \tag{5.5}$$

5.4 Other quantum entropic quantities

By analogy with classical entropies, we can introduce other quantum entropies by considering a composite system AB with associated Hilbert space $\mathcal{H}_A \otimes \mathcal{H}_B$.

Definition 5.4.1. Given the state $\rho_{AB} \in \mathfrak{D}(\mathcal{H}_A \otimes \mathcal{H}_B)$, the quantum joint entropy of AB is defined as

$$S(A, B) \equiv S(\rho_{AB}) := -\mathrm{Tr}(\rho_{AB} \log \rho_{AB}).$$

Definition 5.4.2. Given the state $\rho_{AB} \in \mathfrak{D}(\mathcal{H}_A \otimes \mathcal{H}_B)$, the quantum conditional entropy of A with respect to B is defined as

$$S(A|B) := S(A, B) - S(B).$$

In contrast to the Shannon entropy, the quantum conditional entropy can be negative. Let us illustrate this through an example.

Example. Let $\mathcal{H}_A \simeq \mathcal{H}_B$ and d be its dimension. Take $\rho_{AB} = |\Phi\rangle\langle\Phi|$ with $|\Phi\rangle$, a maximally entangled state vector,

$$|\Phi\rangle = \frac{1}{\sqrt{d}} \sum_i |i\rangle_A |i\rangle_B,$$

where $\{|i\rangle\}_i$ is an orthonormal basis. It then results $\rho_A = \rho_B = \frac{1}{d} I$, therefore we have

$$S(A) = S(\rho_A) = \log d,$$
$$S(B) = S(\rho_B) = \log d,$$

while it is

$$S(A, B) = S(\rho_{AB}) = 0,$$

because ρ_{AB} is pure. Hence, $S(A) \not\leq S(A, B)$, which implies $S(B|A) = S(A, B) - S(A) < 0$.

Definition 5.4.3. Given the state $\rho_{AB} \in \mathfrak{D}(\mathcal{H}_A \otimes \mathcal{H}_B)$, the quantum mutual information between A and B is defined as[1]

$$
\begin{aligned}
I(A;B) &:= S(A) + S(B) - S(A,B) \\
&\equiv S(A) - S(A|B) \\
&\equiv S(B) - S(B|A).
\end{aligned}
$$

5.5 Properties of quantum entropic quantities

We list here the most relevant properties of the above quantities:

(1) Let ρ_A and ρ_B be density operators of $\mathfrak{D}(\mathcal{H}_A)$ and $\mathfrak{D}(\mathcal{H}_B)$, respectively, then

$$
S(\rho_A \otimes \rho_B) \equiv S(A,B) = S(\rho_A) + S(\rho_B) \equiv S(A) + S(B), \quad (5.6)
$$

which is the *additivity property.*

Proof. Let ρ_A and ρ_B have spectral decompositions

$$
\rho_A = \sum_k p_k |\phi_k\rangle\langle\phi_k|, \quad \rho_B = \sum_j q_j |\psi_k\rangle\langle\psi_k|,
$$

then $|\phi_k\rangle|\psi_j\rangle$ is an eigenvector of $\rho_A \otimes \rho_B$ with corresponding eigenvalue $p_k q_j$, hence

$$
\begin{aligned}
S(\rho_A \otimes \rho_B) &= -\sum_{k,j} (p_k q_j) \log(p_k q_j) \\
&= -\left(\sum_j q_j\right) \sum_k p_k \log p_k - \left(\sum_k p_k\right) \sum_j q_j \log q_j \\
&= S(\rho_A) + S(\rho_B). \qquad \square
\end{aligned}
$$

[1] To distinguish more easily, quantum from classical mutual information, we separate the two arguments by ";" instead of ":".

(2) Generally, for a composite system AB it is

$$S(A,B) \leq S(A) + S(B), \tag{5.7}$$

which is the *subadditivity property*, with equality if and only if $\rho_{AB} = \rho_A \otimes \rho_B$.

Proof. According to Theorem 5.2.2, we have

$$\begin{aligned} 0 \leq S(\rho_{AB}||\rho_A \otimes \rho_B) &= \mathrm{Tr}(\rho_{AB} \log \rho_{AB}) - \mathrm{Tr}(\rho_{AB} \log(\rho_A \otimes \rho_B)) \\ &= -S(\rho_{AB}) - \mathrm{Tr}(\rho_{AB} \log(\rho_A \otimes I_B)) \\ &\quad -\mathrm{Tr}(\rho_{AB} \log(I_A \otimes \rho_B)) \\ &= -S(\rho_{AB}) - \mathrm{Tr}(\rho_A \log \rho_A) - \mathrm{Tr}(\rho_B \log \rho_B) \\ &= -S(\rho_{AB}) + S(\rho_A) + S(\rho_B). \end{aligned}$$

Hence, $S(\rho_{AB}) \leq S(\rho_A) + S(\rho_B)$. Note that this also states the non-negativity of the quantum mutual information. □

(3) If a composite system AB is in a pure state, the von Neumann entropy of the subsystems are equal $S(A) = S(B)$.

Proof. It immediately follows from the Schmidt decomposition (3.1). □

(4) Let $\rho = \sum_i p_i \rho_i$ with ρ_i density operators with support on orthogonal subspaces, then

$$S\left(\sum_i p_i \rho_i\right) = H(p) + \sum_i p_i S(\rho_i). \tag{5.8}$$

Proof. Let us consider the spectral decompositions

$$\rho_i = \sum_j \lambda_i^j |e_i^j\rangle\langle e_i^j|,$$

then

$$S\left(\sum_i p_i\rho_i\right) = -\sum_{ij}(p_i\lambda_i^j)\log(p_i\lambda_i^j)$$

$$= -\sum_i p_i \log p_i - \sum_i p_i\left(\sum_j \lambda_i^j \log \lambda_i^j\right)$$

$$= H(p) + \sum_i p_i S(\rho_i).$$

□

(5) The von Neumann entropy is *concave*, i.e.

$$S\left(\sum_i p_i\rho_i\right) \geq \sum_i p_i S(\rho_i). \tag{5.9}$$

Proof. Choose a reference system B with associated $\mathcal{H}_B$ and an orthonormal basis $\{|i\rangle\}_i$, then let

$$\rho^{AB} = \sum_i p_i\rho_i \otimes |i\rangle\langle i|.$$

It follows that

$$S(A) = S\left(\sum_i p_i\rho_i\right)$$

$$S(B) = S\left(\sum_i p_i|i\rangle\langle i|\right) = H(p)$$

$$S(A,B) = H(p) + \sum_i p_i S(\rho_i),$$

due to the fact that $(\rho_i \otimes |i\rangle\langle i|)$ have support on orthogonal subspaces [property (4)]. Then using the subadditivity of von Neumann entropy [property (2)]

$$S(A,B) \leq S(A) + S(B),$$

we have

$$H(p) + \sum_i p_i S(\rho_i) \leq H(p) + S\left(\sum_i p_i\rho_i\right).$$

□

(6) For any state ρ_{ABC} of a tripartite system ABC, it is

$$S(\rho_{ABC}) + S(\rho_B) \leq S(\rho_{AB}) + S(\rho_{BC}), \tag{5.10}$$

which is the *strong subadditivity property.*
The proof of this property goes beyond the scope of this book, hence it is omitted. Let us, however, analyze some consequences of it.

(6.A) Conditioning reduces entropy, i.e.

$$S(A|BC) \leq S(A|B). \tag{5.11}$$

Proof. Starting from strong subadditivity property

$$S(A, B, C) + S(B) \leq S(A, B) + S(B, C),$$

we equivalently get

$$S(A, B, C) - S(B, C) \leq S(A, B) - S(B),$$

which gives the desired inequality. □

(6.B) Discarding quantum systems never increases quantum mutual information, i.e.

$$I(A; B) \leq I(A; B, C). \tag{5.12}$$

Proof. Starting from strong subadditivity property

$$S(A, B, C) + S(B) \leq S(A, B) + S(B, C),$$

we equivalently get

$$-S(A, B) + S(B) + S(A) \leq -S(A, B, C) + S(B, C) + S(A),$$

which gives the desired inequality. □

Exercises

1. Show that, given a (qubit) density operator ρ, a projective measurement performed on it never decreases its von Neumann entropy (consider the mixture of states resulting from all possible measurement outcomes). In contrast, provide an example of POVM that can decrease its von Neumann entropy.

2. Consider the two-qubit state vector $|\psi\rangle_{AB} = \alpha|00\rangle + \beta|11\rangle$ with $\alpha, \beta \in \mathbb{C}$, such that $|\alpha|^2 + |\beta|^2 = 1$. Let $\rho_{AB} = |\psi\rangle_{AB}\langle\psi|$ and find $S(\rho_A)$. Compute the minimum of fidelity $F(\rho_{AB}, \tilde{\rho}_{AB})$ over $\tilde{\rho}_{AB}$, intended as density operator of a separable state. Compare this value with $S(\rho_{AB})$, what is the largest?
3. Show that given two observables A and B on the Hilbert space $\mathcal{H}$ and a concave function, then it is

$$\mathrm{Tr}[f(B) - f(A)] \leq \mathrm{Tr}[(B - A)f'(A)].$$

Use this to show that $S(\rho\|\sigma) \geq 0$, $\forall \rho, \sigma$.
4. Prove the following inequality (due to Araki and Lieb)

$$|S(\rho_1) - S(\rho_2)| \leq S(\rho_{12}).$$

5. Given the quantity $d(\rho, \rho') := 2 - 2F(\rho, \rho')$, show that it is a metric. Then, for $|\psi\rangle = \alpha|00\rangle + \beta|11\rangle$, consider $\rho = |\psi\rangle\langle\psi|$ and ρ' not entangled and find the minimum of $d(\rho, \rho')$ over ρ'.

Chapter 6

CHANNEL MAPS

Information processing in closed classical (resp., quantum) systems is described by permutations on $\mathbb{F}_d$ (resp., by unitaries on $\mathcal{H}_d$) as we have seen, i.e. by reversible maps on the set of states. In this chapter, we shall generalize the description to the case of open systems where the reversibility no longer holds.

6.1 Classical stochastic maps

In the classical framework, a generic process, not necessarily reversible, can be intended as mapping a random variable X on the alphabet $\mathcal{X}$ into another random variable Y on the alphabet $\mathcal{Y}$. Suppose $X \sim p_X(x)$ and $Y \sim p_Y(y)$, then mapping $X \to Y$ can be intended as a map

$$N : \mathfrak{P}(\mathcal{X}) \to \mathfrak{P}(\mathcal{Y}), \tag{6.1}$$

where $\mathfrak{P}(\mathcal{X})$ (resp., $\mathfrak{P}(\mathcal{Y})$) denotes the set of probability distributions over the alphabet $\mathcal{X}$ (resp., $\mathcal{Y}$). Such maps have the following properties:

(i) positivity, i.e.

$$N(p_X) = p_Y \geq 0, \quad \text{for} \quad p_X \geq 0;$$

(ii) preservation of normalization, i.e.

$$\sum_{y \in \mathcal{Y}} N(p_X) = \sum_{y \in \mathcal{Y}} p_Y(y) = 1, \quad \text{for} \quad \sum_{x \in \mathcal{X}} p_X(x) = 1;$$

and in addition we require

(iii) linearity, i.e.

$$N(ap_X + bp_Z) = aN(p_X) + bN(p_Z),$$

for X and Z independent random variables and $a, b \in \mathbb{R}_+$, such that $a + b = 1$.

A stochastic map possesses all these properties. It can be written as

$$p_Y = Np_X, \tag{6.2}$$

where now N is a $|\mathcal{Y}| \times |\mathcal{X}|$ real matrix with each column summing up to 1, known as stochastic matrix (if also $N^\top$ has this property, the matrix is said to be bistochastic). Here, p_X, p_Y are intended as probability (column) vectors.

The entries of a stochastic matrix N can be understood as *conditional probabilities* (or *transition probablities*):

$$N_{yx} = \Pr\{Y = y|X = x\}.$$

Definition 6.1.1. A classical channel is a stochastic map N between the sets of probability distributions $\mathfrak{P}(\mathcal{X})$ and $\mathfrak{P}(\mathcal{Y})$ over two given alphabets, $\mathcal{X}$ and $\mathcal{Y}$.

Actually, a classical channel can be thought of as the restriction of a map $N : \mathbb{R}^{|\mathcal{X}|} \to \mathbb{R}^{|\mathcal{Y}|}$, which is linear, positive and normalization preserving.

Example: The bistochastic matrix over binary alphabet

$$N = \begin{pmatrix} 1-p & p \\ p & 1-p \end{pmatrix}, \tag{6.3}$$

characterizes the *binary symmetric channel*, where $0 \leq p \leq 1$.

Note that a general stochastic matrix is not reversible, that is, its inverse even if it exists as a matrix, is not a stochastic matrix. The only case in which a stochastic matrix is reversible (i.e. its inverse is also a stochastic matrix) is when it is a permutation.

It is easy to show that the convex sum of permutation matrices gives a stochastic matrix. Furthermore, it is possible to prove that a bistochastic matrix can always be written as convex sum of permutations, and that the convex sum of permutations is always a bistochastic matrix.

6.2 Input–output fidelity and data processing inequality

A measure of how well a state is transferred from input to output in a channel is given by the fidelity between input and output probability distributions (see Section 2.3).

A sequence of channels can be seen as a sequence of random variables connected by a stochastic process characterized by a joint probability distribution. A particular case of sequence of random variables is provided by the *Markov chain.*

Definition 6.2.1. A Markov chain is a sequence $X_1 \to X_2 \to \dots$ of random variables, such that X_{n+1} is independent of $X_1 \dots X_{n-1}$ given X_n. That is

$$\Pr\left\{X_{n+1} = x_{n+1} \middle| X_1 = x_1, \dots, X_n = x_n\right\}$$
$$= \Pr\left\{X_{n+1} = x_{n+1} \middle| X_n = x_n\right\}.$$

Theorem 6.2.2 (Data processing inequality). *Given a Markov chain $X \to Y \to Z$, it follows that*

$$I(X : Y) \geq I(X : Z).$$

Proof. We have

$$\begin{aligned} I(X : Y, Z) &= I(X : Z) + I(X : Y|Z), \\ &= I(X : Y) + I(X : Z|Y). \end{aligned}$$

Also, $I(X : Z|Y) = 0$, since X and Z are conditionally independent, hence, we can write

$$I(X : Z) + I(X : Y|Z) = I(X : Y). \tag{6.4}$$

Now we may note that

$$I(X : Y|Z) = H(X|Z) - H(X|Y, Z) \geq 0,$$

where the inequality comes from the fact that conditioning reduces entropy, i.e. $H(X|Y,Z) \leq H(X|Z)$. Finally, using $I(X:Y|Z) \geq 0$ in (6.4), we get the desired result. $\square$

The data processing inequality states that any further manipulation of data cannot be used to increase the amount of mutual information between the final and the original variables.

6.3 Quantum stochastic maps

In analogy to the classical case, we are going to consider a map

$$\mathcal{N} : \mathfrak{D}(\mathcal{H}_A) \to \mathfrak{D}(\mathcal{H}_B), \tag{6.5}$$

as candidate to describe a quantum channel.

Such a map must satisfy the following properties:

(i) positivity, i.e.

$$\mathcal{N}(\rho) = \rho' \geq 0;$$

(ii) preservation of normalization (or trace), i.e.

$$\mathrm{Tr}\,(\mathcal{N}(\rho)) = \mathrm{Tr}\,\rho' = 1.$$

In addition, we require

(iii) complete positivity, i.e.

$$(\mathcal{N} \otimes \mathrm{id}_R)(\sigma) \geq 0,$$

for any $\sigma \in \mathfrak{D}(\mathcal{H}_A \otimes \mathcal{H}_R)$, whatever is the extension of the map $\mathcal{N}$ (i.e. whatever is the reference system R).

(iv) linearity, i.e.

$$\mathcal{N}(a\rho_1 + b\rho_2) = a\mathcal{N}(\rho_1) + b\mathcal{N}(\rho_2),$$

for $\rho_1, \rho_2 \in \mathfrak{D}(\mathcal{H}_A)$ and $a, b \in \mathbb{R}_+$, such that $a + b = 1$.

These properties are analogous to those of classical stochastic maps. The only exception is (iii), which is stronger than the simple positivity. Its requirement arises from the fact that given $\rho_A \in \mathfrak{D}(\mathcal{H}_A)$, we know that it can be purified into $|\psi\rangle_\rho\langle\psi| \in \mathfrak{D}(\mathcal{H}_A \otimes \mathcal{H}_R)$ with R as a reference system (Theorem 4.3.2). Then applying the map $(\mathcal{N} \otimes \mathrm{id}_R)$ to such a state, we must have a valid density operator, i.e. $(\mathcal{N} \otimes \mathrm{id}_R)(|\psi\rangle_\rho\langle\psi|) \geq 0$.

Definition 6.3.1. A quantum channel is a linear map

$$\mathcal{N} : \mathfrak{D}(\mathcal{H}_A) \to \mathfrak{D}(\mathcal{H}_B),$$

which is also completely positive.

Actually, a quantum channel can be thought of as the restriction of a map $\mathcal{N} : \mathfrak{L}(\mathcal{H}_A) \to \mathfrak{L}(\mathcal{H}_B)$, which is linear, completely positive and trace preserving (CPTP).

We have the following important result about the representation of quantum channels.

Theorem 6.3.2 (Kraus representation). *A map $\mathcal{N} : \mathfrak{L}(\mathcal{H}_A) \to \mathfrak{L}(\mathcal{H}_B)$ is linear and CPTP if and only if it can be written as*

$$\mathcal{N}(\rho) = \sum_k A_k \rho A_k^\dagger,$$

for some set of operator $\{A_k\}_k$, which map $\mathcal{H}_A$ into $\mathcal{H}_B$ and such that $\sum_k A_k^\dagger A_k = I$.

Proof. We first prove that any $\mathcal{N}$ written in the Kraus form is linear and CPTP. Linearity is evident. To demonstrate CP, it suffices to show

$$\langle\psi| (\mathcal{N} \otimes \mathrm{id}) (\sigma) |\psi\rangle \geq 0,$$

for any $|\psi\rangle \in \mathcal{H}_A \otimes \mathcal{H}_R$ and for any positive operator $\sigma \in \mathfrak{L}(\mathcal{H}_A \otimes \mathcal{H}_R)$. By Kraus representation, we have

$$\sum_k \langle\psi| (A_k \otimes I) (\sigma) \left(A_k^\dagger \otimes I\right) |\psi\rangle = \sum_k \langle\phi_k|\sigma|\phi_k\rangle,$$

where $|\phi_k\rangle := (A_k^\dagger \otimes I)|\psi\rangle$. Since σ is positive, the desired result follows immediately. Furthermore, $\sum_k A_k^\dagger A_k = I$ implies the preservation of trace.

Let us now prove that any linear CPTP map can be written in the Kraus form. Define

$$|\psi\rangle_{AR} := \frac{1}{\sqrt{d}} \sum_{i=1}^{d} |i_A\rangle|i_R\rangle \in \mathcal{H}_A \otimes \mathcal{H}_R,$$

where $d := |\mathcal{H}_A| = |\mathcal{H}_R|$. Also define

$$|\tilde{\psi}\rangle_{AR} := \sqrt{d}|\psi\rangle_{AR} = \sum_{i=1}^{d} |i_A\rangle|i_R\rangle, \tag{6.6}$$

$$|\varphi\rangle_A := \sum_i a_i |i\rangle_A \in \mathcal{H}_A, \qquad a_i \in \mathbb{C}, \tag{6.7}$$

$$|\varphi^*\rangle_R := \sum_i a_i^* |i\rangle_R \in \mathcal{H}_R, \qquad a_i \in \mathbb{C}. \tag{6.8}$$

Then we have the following "partial inner product"

$${}_R\langle\varphi^*|\tilde{\psi}\rangle_{AR} = |\varphi\rangle_A. \tag{6.9}$$

Hence, it is

$$\begin{aligned} \mathcal{N}\left(|\varphi\rangle_A\langle\varphi|\right) &= \mathcal{N}\left({}_R\langle\varphi^*|\tilde{\psi}\rangle_{AR}\langle\tilde{\psi}|\varphi^*\rangle_R\right) \\ &= {}_R\langle\varphi^*|\left(\mathcal{N}\otimes \mathrm{id}_R\right)\left(\tilde{\rho}_{AR}\right)|\varphi^*\rangle_R, \end{aligned}$$

where $\tilde{\rho}_{AR} := |\tilde{\psi}\rangle_{AR}\langle\tilde{\psi}|$. Since $\mathcal{N}$ is a linear CPTP map, the action of $(\mathcal{N}\otimes \mathrm{id}_R)$ on $\tilde{\rho}_{AR}$ will give a density operator on $\mathfrak{D}(\mathcal{H}_B \otimes \mathcal{H}_R)$, say $\tilde{\rho}'_{BR}$. Then we can write

$$\mathcal{N}\left(|\varphi\rangle_A\langle\varphi|\right) = {}_R\langle\varphi^*|\tilde{\rho}'_{BR}|\varphi^*\rangle_R.$$

However, the density operator $\tilde{\rho}'_{BR}$ can be associated to an ensemble of pure states $\{\lambda_k, |\tilde{\alpha}_k\rangle_{BR}\}_k$ (which also provides its spectral decomposition), thus

$$\mathcal{N}\left(|\varphi\rangle_A\langle\varphi|\right) = \sum_k \lambda_k \langle\varphi_R^*|\tilde{\alpha}_k\rangle_{BR}\langle\tilde{\alpha}_k|\varphi_R^*\rangle.$$

Now define the operator $A_k : \mathcal{H}_A \to \mathcal{H}_B$ by means of

$$A_k\left(|\varphi\rangle_A\right) = \sqrt{\lambda_k}\,{}_R\langle\varphi^*|\tilde{\alpha}_k\rangle_{BR}.$$

It results a linear operator due to the linearity of the "partial inner product" (6.9). As a consequence, we have

$$\mathcal{N}\left(|\varphi\rangle_A\langle\varphi|\right) = \sum_k A_k|\varphi\rangle_A\langle\varphi|A_k^\dagger.$$

Furthermore, as $\mathcal{N}$ is trace preserving, we also have $\sum_k A_k^\dagger A_k = I$. □

It is worth remarking that since $\tilde{\rho}'_{AR}$ has a rank at most d^2 (it has at most d^2 non-zero eigenvalues), $\mathcal{N}$ possesses a Kraus representation with at most d^2 Kraus operators.

Corollary 6.3.3. *Suppose* $\{A_1, \ldots, A_m\}$ *and* $\{B_1, \ldots, B_n\}$ *are Kraus operators giving rise to the map* $\mathcal{N}$ *and* $\mathcal{N}'$*, respectively. By appending zero operators to the shorter list, we may ensure that* $m = n$*. Then,* $\mathcal{N} = \mathcal{N}'$ *if and only if there exists complex numbers* u_{ij} *such that* $A_i = \sum_j u_{ij} B_j$ *with* u_{ij} *entries of a unitary transformation.*

The Kraus representation can be thought of as coming from a unitary (closed) evolution between the main system and an "environment" by disregarding the latter at the end, that is

$$\mathcal{N}(\rho) = \mathrm{Tr}_E\left[U\left(\rho \otimes \rho_E\right)U^\dagger\right]. \tag{6.10}$$

In fact, by assuming $\rho_E = |e_0\rangle\langle e_0|$, we have[1]

$$\mathcal{N}(\rho) = \sum_k \langle e_k|\left[U\left(\rho \otimes |e_0\rangle\langle e_0|\right)U^\dagger\right]|e_k\rangle,$$

where $\{|e_k\rangle\}_k$ is an orthonormal basis. Therefore, we can write $\mathcal{N}(\rho) = \sum_k A_k \rho A_k^\dagger$ with $A_k := \langle e_k|U|e_0\rangle$.

The unitary U is called *(Stinespring) dilation of* $\mathcal{N}$. Since the initial state of the environment $|e_0\rangle$ is fixed, it can be effectively considered as an isometry $V^{A\to BE}$ defined as $V|\psi\rangle := U(|\psi\rangle \otimes |e_0\rangle)$.

[1]If ρ_E is not pure, we can always think to purify it by considering a larger environment E, according to Theorem 4.3.2.

Definition 6.3.4. Given (6.10), one can also define the quantum channel $\widetilde{\mathcal{N}}$ complementary to $\mathcal{N}$ as

$$\widetilde{\mathcal{N}}(\rho) := \mathrm{Tr}_B\left[U(\rho\otimes\rho_E)U^\dagger\right]. \tag{6.11}$$

Definition 6.3.5. A quantum channel $\mathcal{N}$ that leaves invariant the maximally mixed state, i.e. $\mathcal{N}\left(\frac{1}{d}I\right)=\frac{1}{d}I$, is called unital.

Definition 6.3.6. Two quantum channels $\mathcal{M}$ and $\mathcal{N}$ are said to be unitarily equivalent if they are related by unitary channels on input and output, i.e. if

$$\mathcal{M}(\rho) = U\mathcal{N}(V\rho V^\dagger)U^\dagger,$$

for some unitary transformations U and V.

Definition 6.3.7. A quantum channel $\mathcal{N}$ is called entanglement-breaking if $(\mathcal{N}_A\otimes \mathrm{id}_R)\sigma_{AR}$ gives a separable state for any bipartite input σ_{AR} (thus, including entangled states) on the main system A and the reference system R.

6.4 Quantum input–output fidelity and data processing inequality

A measure of how well a quantum state is transferred through a quantum channel is given by the fidelity between the input state ρ and the corresponding output state $\mathcal{N}(\rho)$ (see Section 5.3). However, we know (by Theorem 4.3.2) that given a mixed state ρ_Q for the system Q, it can always be regarded as entangled with a reference system R that purifies it, namely $\rho_Q = \mathrm{Tr}_R(|\psi\rangle_{QR}\langle\psi|)$. Then we could ask how well is the entanglement between systems Q and R preserved by the quantum channel $\mathcal{N}$ acting only on the system Q? To this end we have to consider the fidelity between the states $|\psi\rangle_{QR}\langle\psi|$ and $(\mathcal{N}\otimes \mathrm{id}_R)(|\psi\rangle_{QR}\langle\psi|)$, that is

$$F(\rho,\mathcal{N}) := {}_{QR}\langle\psi|(\mathcal{N}\otimes \mathrm{id}_R)(|\psi\rangle_{QR}\langle\psi|)|\psi\rangle_{QR}. \tag{6.12}$$

This fidelity is known as *entanglement fidelity.* Given $\mathcal{N}(\rho) = \sum_k A_k \rho A_k^\dagger$, it results

$$F(\rho, \mathcal{N}) = \sum_k |\mathrm{Tr}\,(\rho A_k)|^2 . \tag{6.13}$$

An important property of quantum channels is that the distance coming from fidelity (see Section 5.3) is *contractive.*

Theorem 6.4.1. *Given two density operators acting on the Hilbert space $\mathcal{H}$,*

$$d\,(\mathcal{N}(\rho), \mathcal{N}(\sigma)) \leq d(\rho, \sigma),$$

or equivalently

$$F\,(\mathcal{N}(\rho), \mathcal{N}(\sigma)) \geq F(\rho, \sigma).$$

Proof. Let $|\psi_\rho\rangle$ and $|\varphi_\sigma\rangle$ be the purifications of ρ and σ, respectively, such that $F(\rho, \sigma) = |\langle \psi_\rho | \varphi_\sigma \rangle|$. Introduce an environment for $\mathcal{N}$ which is in the state $|e_0\rangle$, then $U|\psi_\rho\rangle|e_0\rangle$ results as a purification of $\mathcal{N}(\rho)$ and analogously $U|\varphi_\rho\rangle|e_0\rangle$ a purification of $\mathcal{N}(\sigma)$. Hence, as a consequence of the definition (5.3.1), we have

$$\begin{aligned} F\,(\mathcal{N}(\rho), \mathcal{N}(\sigma)) &\geq \left| \langle \psi_\rho | \langle e_0 | U^\dagger U | \varphi_\sigma \rangle | e_0 \rangle \right| \\ &\geq |\langle \psi_\rho | \varphi_\sigma \rangle| = F(\rho, \sigma). \end{aligned}$$

□

The contractive property states that a quantum channel makes two states less distinguishable. Another important property of quantum channels is that mutual information never increases under their action.

Theorem 6.4.2. *Suppose AB labels a composite system and $\mathcal{N}$ is a quantum channel acting only on the subsystem B. Let $I(A; B)$ (resp., $I(A'; B')$) be the mutual information before (resp., after) the action of the channel $\mathrm{id}_A \otimes \mathcal{N}$, then*

$$I(A'; B') \leq I(A; B).$$

Proof. According to the Stinespring dilation, the action of $\mathcal{N}$ on B may be simulated by introducing an environment E in a pure state $|e_0\rangle\langle e_0|$ and a unitary U_{BE}. The action of $\mathcal{N}$ then comes from such U_{BE} by discarding E. So prior to the action of the channel it is

$$I(A;B) = I(A;B,E),$$

because the global system is in a product state between AB and E and the latter is in a pure state. Since U_{BE} does not affect the subsystem A, we will then have

$$I(A;B,E) = I(A';B',E'),$$

due to the invariance of (joint) von Neumann entropy. Now according to Section 5.5, discarding a system cannot increase mutual information, hence we end up with

$$I(A';B') \leq I(A';B',E') = I(A;B). \qquad \square$$

Quantum mutual information also obeys subadditivity.

Theorem 6.4.3. *Consider two channels $\mathcal{N}_1$ and $\mathcal{N}_2$ with input Q_1 and Q_2 coupled, respectively, with environment E_1 and E_2. Furthermore, let R be the reference system purifying the joint system Q_1Q_2. Then we have*

$$I(R;Q_1'Q_2') \leq I(RQ_1;Q_2') + I(RQ_2;Q_1'),$$

where prime denotes system after channel action.

Proof. It holds

$$\begin{aligned} S(Q_1'Q_2'|E_1'E_2') &= S(Q_1'|E_1'E_2') + S(Q_2'|E_1'E_2') - I(Q_1';Q_2'|E_1'E_2') \\ &\leq S(Q_1'|E_1'E_2') + S(Q_2'|E_1'E_2'), \qquad (6.14) \\ &\leq S(Q_1'|E_1') - I(Q_1';E_2'|E_1') + S(Q_2'|E_2') \\ &\quad -I(Q_2';E_1'|E_2') \leq S(Q_1'|E_1') + S(Q_2'|E_2'). \quad (6.15) \end{aligned}$$

The terms ignored in obtaining (6.14) and (6.15) are non-negative due to strong subadditivity (Section 5.5). For instance, $I(Q_1';E_2'|E_1')$ can be written as $S(Q_1') + S(E_2') + S(E_1') - S(Q_1',E_2',E_1') - I(Q_1';E_1') - I(E_2';E_1')$. Then writing out the mutual informations

$I(Q_1'; E_1')$ and $I(E_2'; E_1')$, and applying the strong subadditivity, yields $I(Q_1'; E_2'|E_1') \geq 0$.

Using (6.15) together with subadditivity of ordinary von Neumann entropy leads to the desired inequality. $\square$

To find out the quantum analog of the classical data processing inequality (see Theorem 6.2.2), we need to introduce a new entropic quantity.

Let R denote a reference system for purifying a state ρ of the system Q, i.e. $\rho = \mathrm{Tr}_R(|\psi\rangle_{RQ}\langle\psi|)$. Due to the action of a channel $\mathcal{N}$ on Q, the composite system RQ evolves into $\rho'_{RQ} = (\mathrm{id}_R \otimes \mathcal{N})\rho_{RQ}$ and the final states of the system Q will be $\rho'_Q = \mathrm{Tr}_R(\rho'_{RQ})$. We know that $\mathcal{N}$ can be regarded as coming from a unitary acting on a larger system, thus let us introduce an environment E in a fixed initial states, $|e_0\rangle\langle e_0|$. Then

$$\rho'_Q = \mathrm{Tr}_E \left[U_{QE} \left(\rho_Q \otimes |e_0\rangle\langle e_0| \right) U_{QE}^\dagger \right].$$

The initial state of the system RQE is $|\psi\rangle_{RQE} = |\psi_\rho\rangle_{RQ}|e_0\rangle$, while the final (still pure) reads $|\psi'\rangle_{RQE} = (I_R \otimes U_{QE})|\psi\rangle_{RQE}$. Let $\rho_{RQE} := |\psi\rangle_{RQE}\langle\psi|$ and $\rho'_{RQE} := |\psi'\rangle_{RQE}\langle\psi'|$.

Definition 6.4.4. The coherent information of a quantum channel $\mathcal{N}$ with respect to the input state ρ_Q is defined as

$$\begin{aligned} I_{\mathrm{coh}}(\rho, \mathcal{N}) &:= S(\rho'_Q) - S(\rho'_{RQ}) \\ &\equiv S(\rho'_Q) - S(\rho'_E) \equiv S(\rho'_{RE}) - S(\rho'_E). \end{aligned}$$

Note that the equivalence among the definitions comes from the property that the entropy of subsystems are equal if the global system is in a pure state (see Section 5.5).

The coherent information can be positive, negative or zero. It can be thought as a measure of how "non-classical" the final state ρ'_{RQ} is. The initial state vector $|\psi_\rho\rangle_{RQ}$ can be entangled and $I_{\mathrm{coh}}(\rho, \mathcal{N})$ measures the degree to which the entanglement is retained between R and Q after the action of the channel (if $|\psi_\rho\rangle_{RQ}$ is not entangled $I_{\mathrm{coh}} = 0$).

Using the subadditivity of von Neumann entropy (see Section 5.5) we have

$$\begin{aligned} I_{\mathrm{coh}}(\rho, \mathcal{N}) &= S(\rho'_{RE}) - S(\rho'_E) \\ &\leq S(\rho'_R) + S(\rho'_E) - S(\rho'_E) = S(\rho'_R), \end{aligned}$$

however $S(\rho'_R) = S(\rho_R)$ since R is not affected by $\mathcal{N}$. Hence, we have

$$I_{\mathrm{coh}}(\rho, \mathcal{N}) \leq S(\rho), \tag{6.16}$$

using the fact that $|\psi^{\rho}_{RQ}\rangle\langle\psi^{\rho}_{RQ}|$ is pure.

Theorem 6.4.5 (Quantum data processing inequality). *Let us consider the following process:*

$$\rho \to \mathcal{N}_1(\rho) \to \mathcal{N}_2 \circ \mathcal{N}_1(\rho) \equiv \mathcal{N}_2(\mathcal{N}_1(\rho)),$$

then

$$S(\rho) \geq I_{coh}(\rho, \mathcal{N}_1) \geq I_{coh}(\rho, \mathcal{N}_2 \circ \mathcal{N}_1).$$

Proof. Consider $|\psi\rangle_{RQE_1E_2} = |\psi_\rho\rangle_{RQ}|0\rangle_{E_1}|0\rangle_{E_2}$. After the action of the unitary U_{QE_1} charaterizing $\mathcal{N}_1$, we have

$$\begin{aligned} |\psi'\rangle_{RQE_1E_2} &= (I_R \otimes U_{QE_1} \otimes I_{E_2})\,|\psi\rangle_{RQE_1E_2} \\ &= |\psi'\rangle_{RQE_1}|0\rangle_{E_2}. \end{aligned}$$

Let $\rho'_{RQE_1E_2} := |\psi'\rangle_{RQE_1E_2}\langle\psi'|$. After the action of the unitary U_{QE_2} charaterizing $\mathcal{N}_2$, we have

$$|\psi''\rangle_{RQE_1E_2} = (I_R \otimes I_{E_1} \otimes U_{QE_2})\,|\psi'\rangle_{RQE_1E_2}.$$

Let $\rho''_{RQE_1E_2} := |\psi''\rangle_{RQE_1E_2}\langle\psi''|$.

Now let us apply the strong subadditivity property of von Neumann entropy (see Section 5.5) to the final state of the composite system RE_1E_2. We will get

$$S(\rho''_{RE_1E_2}) + S(\rho''_{E_1}) \leq S(\rho''_{RE_1}) + S(\rho''_{E_1E_2}). \tag{6.17}$$

Note that $S(\rho''_{RE_1E_2}) = S(\rho''_Q)$ since $\rho''_{RQE_1E_2}$ is pure. Furthermore, it is $S(\rho''_{E_1}) = S(\rho'_{E_1})$, and $S(\rho''_{RE_1}) = S(\rho'_{RE_1})$, since R and E_1 are not affected by $\mathcal{N}_2$. Also, $S(\rho''_{RE_1}) = S(\rho'_{RE_1}) = S(\rho'_Q)$ because ρ'_{RQE_1} is pure. With all that just considered, we can rewrite (6.17) as

$$S(\rho''_Q) + S(\rho'_{E_1}) \leq S(\rho'_Q) + S(\rho''_{E_1E_2}),$$

or

$$S(\rho''_Q) - S(\rho''_{E_1E_2}) \leq S(\rho'_Q) - S(\rho'_{E_1}).$$

□

Corollary 6.4.6 (Data processing inequality for quantum mutual information). *Let us consider the following process:*

$$\begin{aligned}\rho_{RQ} \to (\mathrm{id} \otimes \mathcal{N}_1)(\rho_{RQ}) &\to (\mathrm{id} \otimes \mathcal{N}_2) \circ (\mathrm{id} \otimes \mathcal{N}_1)(\rho_{RQ}) \\ &\equiv (\mathrm{id} \otimes \mathcal{N}_2)\left((\mathrm{id} \otimes \mathcal{N}_1)(\rho_{RQ})\right),\end{aligned}$$

then

$$I(R;Q) \geq I(R;Q_1) \geq I(R;Q_2).$$

Proof. The proof follows from the use of Theorem 6.4.2 and from the fact that $I(R;Q_1) = S(R) + I_{\mathrm{coh}}(\rho, \mathcal{N})$ (see Definition 6.4.4) by applying the quantum data processing inequality for coherent information. □

6.5 A survey of qubit channels

A compact characterization of qubit quantum channels, i.e. linear CPTP maps from $\mathfrak{D}(\mathbb{C}^2)$ to itself, can be obtained by adopting the Bloch ball representation (see Postulate 4.2.1) according to which a qubit density operator is uniquely identified by a vector $\vec{r} = (r_x, r_y, r_z) \in \mathbb{R}^3$ such that $r_x^2 + r_y^2 + r_z^2 \leq 1$

$$\rho = \frac{1}{2}\begin{pmatrix} 1 + r_z & r_x - ir_y \\ r_x + ir_y & 1 - r_z \end{pmatrix} = \frac{1}{2}\left(1 + \sigma_{\vec{r}}\right).$$

In this framework, any qubit channel $\mathcal{N}$ induces an *affine* transformation of the type

$$\vec{r} \overset{\mathcal{N}}{\longmapsto} \vec{r}\,' = M\vec{r} + \vec{t},$$

where $M \in \mathrm{GL}(3,\mathbb{R})$ and $\vec{t} = (t_x, t_y, t_z) \in \mathbb{R}^3$. In particular, qubit unital channels are obtained for $\vec{t} = 0$ and $M^\top M \leq I$ (with equality satisfied only for unitary channels).

- **Bit-flip channel.** It maps $|0\rangle \to |1\rangle$ and $|1\rangle \to |0\rangle$ probabilistically. The Kraus operators (in the canonical basis) are

$$A_0 = \sqrt{1-p}\begin{pmatrix} 1 & 0 \\ 0 & 1 \end{pmatrix}, \quad A_1 = \sqrt{p}\begin{pmatrix} 0 & 1 \\ 1 & 0 \end{pmatrix}. \tag{6.18}$$

Then a state ρ changes according to

$$\rho \to \mathcal{N}(\rho) = (1-p)\rho + p\,\sigma^X \rho \sigma^X. \tag{6.19}$$

The interpretation is that the state is left unchanged with probability $1-p$, and a bit-flip (in the canonical basis) happens with probability p. This channel maps the Bloch vector as follows: $(r_x, r_y, r_z) \to (r_x, (1-2p)r_y, (1-2p)r_z)$.

- **Phase-flip channel.** It maps $|0\rangle \to |0\rangle$ and $|1\rangle \to -|1\rangle$ probabilistically. The Kraus operators (in the canonical basis) are

$$A_0 = \sqrt{1-p}\begin{pmatrix} 1 & 0 \\ 0 & 1 \end{pmatrix}, \quad A_1 = \sqrt{p}\begin{pmatrix} 1 & 0 \\ 0 & -1 \end{pmatrix}. \tag{6.20}$$

Then a state ρ changes according to

$$\rho \to \mathcal{N}(\rho) = (1-p)\rho + p\,\sigma^Z \rho \sigma^Z. \tag{6.21}$$

The interpretation is that the state is left unchanged with probability $(1-p)$, and a phase-flip (in the canonical basis) happens with probability p. The transformation on the Bloch vector is $(r_x, r_y, r_z) \to ((1-2p)r_x, (1-2p)r_y, r_z)$.

- **Bit-phase-flip channel.** It is the combination of a bit-flip and a phase-flip. By noticing that $\sigma^Y = i\sigma^X\sigma^Z$, it is represented by the

Kraus operators (in the canonical basis)

$$A_0 = \sqrt{1-p}\begin{pmatrix} 1 & 0 \\ 0 & 1 \end{pmatrix}, \quad A_1 = \sqrt{p}\begin{pmatrix} 0 & -i \\ i & 0 \end{pmatrix}. \tag{6.22}$$

Then a state ρ changes according to

$$\rho \to \mathcal{N}(\rho) = (1-p)\rho + p\,\sigma^Y \rho \sigma^Y. \tag{6.23}$$

The interpretation is that the state is left unchanged with probability $(1-p)$, and a phase-flip and bit-flip happen (in the canonical basis) with probability p. The transformation on the Bloch vector is $(r_x, r_y, r_z) \to ((1-2p)r_x, r_y, (1-2p)r_z)$.

- **Depolarizing channel.** Loosely speaking, *polarization* means that a vector points towards a well-defined direction. If this direction is randomized, then the vector undergoes *depolarization.* For vectors representing quantum states, this process is described by the following channel map:

$$\mathcal{N}(\rho) = (1-p)\rho + p\frac{I}{2}. \tag{6.24}$$

Hence, we can say that the state is left unchanged with probability $(1-p)$ and it is completely depolarized with probability p. The transformation on the Bloch vector is $(r_x, r_y, r_z) \to ((1-p)r_x, (1-p)r_y, (1-p)r_z)$. The Kraus decomposition can be obtained by noticing that

$$\frac{I}{2} = \frac{\rho + \sigma^X \rho \sigma^X + \sigma^Y \rho \sigma^Y + \sigma^Z \rho \sigma^Z}{4},$$

which implies

$$\mathcal{N}(\rho) = \left(1 - \frac{3p}{4}\right)\rho + \frac{p}{4}\left(\sigma^X \rho \sigma^X + \sigma^Y \rho \sigma^Y + \sigma^Z \rho \sigma^Z\right). \tag{6.25}$$

Note that with the latter representation, we can abandon the original interpretation of p as a probability, because Eq. (6.25) defines a bona fide CPTP map if and only if $0 \le p \le \frac{4}{3}$.

- **Pauli channels** generalize the depolarizing channel, allowing for an arbitrary probability distribution (p_0, p_X, p_Y, p_Z) over Pauli matrices in (6.25):

$$\mathcal{N}(\rho) = p_0\rho + p_X\sigma^X\rho\sigma^X + p_Y\sigma^Y\rho\sigma^Y + p_Z\sigma^Z\rho\sigma^Z, \tag{6.26}$$

which defines a CPTP map iff $p_i \geq 0$ and $p_0 + p_X + p_Y + p_Z = 1$.

- **Amplitude damping channel.** This channel models the decay from one state, e.g. $|0\rangle$, of the canonical basis to the other, e.g. $|1\rangle$. The Kraus operators are

$$A_0 = \begin{pmatrix} \sqrt{1-p} & 0 \\ 0 & 1 \end{pmatrix}, \quad A_1 = \begin{pmatrix} 0 & 0 \\ \sqrt{p} & 0 \end{pmatrix}. \tag{6.27}$$

Then a state ρ changes according to

$$\frac{1}{2}\begin{pmatrix} 1+r_z & r_x - ir_y \\ r_x + ir_y & 1-r_z \end{pmatrix}$$
$$\to \mathcal{N}(\rho) = \frac{1}{2}\begin{pmatrix} (1+r_z)(1-p) & (r_x - ir_y)\sqrt{1-p} \\ (r_x + ir_y)\sqrt{1-p} & 1-r_z+r_z p \end{pmatrix}. \tag{6.28}$$

The transformation on the Bloch vector is $(r_x, r_y, r_z) \to (\sqrt{1-p}r_x, \sqrt{1-p}r_y, (1-p)r_z - p)$. Notice that this channel is not unital (geometrically it does not preserve the center of the Bloch ball, corresponding to the maximally mixed state, $I/2$).

- **Erasure channel.** It describes a communication scenario where errors are somehow heralded. It can be modeled by a CPTP linear map $\mathcal{N} : \mathfrak{D}(\mathbb{C}^2) \to \mathfrak{D}(\mathbb{C}^3)$ acting as

$$\mathcal{N}(\rho) = (1-p)\rho + p|e\rangle\langle e|, \tag{6.29}$$

where $|e\rangle \in \mathbb{C}^3 \ominus \mathbb{C}^2$ (the orthogonal complement of $\mathbb{C}^2$ with respect to $\mathbb{C}^3$). Actually, $|e\rangle$ is a flag to indicate that information has been erased. The Karus operators are as follows:

$$A_0 = \sqrt{1-p}\begin{pmatrix} 1 & 0 \\ 0 & 1 \\ 0 & 0 \end{pmatrix}, \quad A_1 = \sqrt{p}\begin{pmatrix} 0 & 0 \\ 0 & 0 \\ 1 & 0 \end{pmatrix},$$

$$A_2 = \sqrt{p}\begin{pmatrix} 0 & 0 \\ 0 & 0 \\ 0 & 1 \end{pmatrix}. \tag{6.30}$$

Exercises

1. Given a depolarizing quantum channel in $\mathfrak{D}(\mathbb{C}^d)$,

$$\mathcal{N}(\rho) = \lambda\rho + (1-\lambda)\frac{I}{d},$$

determine the range of λ in which the transpose d-depolarizing quantum transformation

$$\rho \mapsto \lambda\rho^{\top} + (1-\lambda)\frac{I}{d}$$

defines a CPTP map.

2. Let $\mathcal{N}$ be the qubit amplitude damping channel. Determine for which value of the parameter p the channel breaks the entanglement between input and reference system R. Draw some conclusions about the form of Kraus operators for *entanglement breaking* quantum channels.

3. Suppose we have a four-dimensional state vectors space. Let us denote by $\{|1\rangle, |2\rangle, |3\rangle, |4\rangle\}$ an orthonormal basis and by P_{12} (resp. P_{34}) the projector onto the subspace spanned by $|1\rangle, |2\rangle$ (resp. $|3\rangle, |4\rangle$). Let U be a unitary operation defined by

$$U = |1\rangle\langle 3| + |2\rangle\langle 4| + |3\rangle\langle 1| + |4\rangle\langle 2|.$$

The channel action is defined by

$$\mathcal{N}(\rho) = P_{12}\,\rho\,P_{12} + UP_{34}\,\rho\,P_{34}U^{\dagger}.$$

Use the map

$$\mathcal{E}(\rho) = \frac{1}{2}P_{12}\,\rho\,P_{12} + \frac{1}{2}U^{\dagger}P_{34}\,\rho\,P_{34}U + P_{34}\,\rho\,P_{34},$$

to show that $I_{\mathrm{coh}}(\rho, \mathcal{N}\circ\mathcal{E}) \leq I_{\mathrm{coh}}(\mathcal{E}(\rho), \mathcal{N})$.

4. Show that any unital qubit channel is unitarily equivalent to a Pauli channel.

5. An (asymmetric) Pauli cloning machine (PCM) is a CPTP linear map $\mathcal{T}$ with a qubit, X, as input and two qubits, A and B as outputs, in such a way that both $\mathrm{Tr}_A \circ \mathcal{T}$ and $\mathrm{Tr}_B \circ \mathcal{T}$ are Pauli channels, from X to B and from X to A, respectively. We can describe a PCM by its unitary dilation U, which acts on three

qubits X, A and B, the latter two initialised in $|0\rangle$, and mapping to three qubits A, B and C; the third qubit C is traced out. Consider a reference qubit R, initially in the maximally entangled state $|\Phi^+\rangle$ with X, and the overall map $I_R \otimes U$ on the initial pure state results in a four-qubit state $|\psi\rangle$ on R and A, B, C.

(i) Show that $\rho_{AR} = \mathrm{Tr}_{BC}|\psi\rangle\langle\psi|$ and $\rho_{BR} = \mathrm{Tr}_{AC}|\psi\rangle\langle\psi|$ are convex mixtures of Bell states.

(ii) Show that indeed any pair among R, A, B, C is in a reduced state that is a mixture of Bell states.

(iii) Write down the action of the PCM on a pure state $|\phi\rangle\langle\phi|$ of the qubit X.

(iv) Show that the optimal symmetric version of PCM, i.e. the one for which the outputs A and B emerge from identical Pauli channels, achieves a fidelity of 5/6.

Chapter 7

INTERLUDE: ESTIMATION THEORY

In the following chapters, we shall deal with the theory of communication and always assume *a priori* knowledge of the property of a channel. It is important, therefore, to know how to estimate the channel properties. This chapter represents a brief interlude devoted to estimation theory. It actually focuses on the estimation of a parameter (which can be then intended as characterizing a channel).

7.1 Classical estimation

A standard problem in estimation theory is to determine the parameter(s) of a distribution from a sample of data drawn from that distribution. For example, let $X_1, \ldots, X_n$ be drawn independently and identically distributed (i.i.d.) according to $p(0) = 1-\theta, p(1) = \theta$, where $\theta \in [0, 1]$ is the parameter to be estimated. As estimator, we could use the first sample X_1. Although its expected value is θ, it is clear that we can do better by using more data, for instance, by suing the sample mean $\frac{1}{n}\sum_i X_i$.

Let X be a random variable on $\mathcal{X}$ with distribution $f(x;\theta)$, which depends continuously on an unknown parameter $\theta \in \mathcal{O}$, the latter being the parameter set.[1] Let $(X_1, \ldots, X_n)$ be the random sample of X. The joint probability distribution is then

$$f(x_1, \ldots, x_n; \theta) = \prod_{i=1}^{n} f(x_i; \theta), \tag{7.1}$$

[1]Here, $\mathcal{O}$ is assumed to be a smooth manifold.

where $x_1, \ldots, x_n$ are the values of the observed data taken from the random sample.

Definition 7.1.1. An estimator of θ for a sample of size n is a function $\hat{\Theta} : \mathcal{X}^n \to \mathcal{O}$. For a particular set of observations $X_1 = x_1, \ldots X_n = x_n$, the value of the estimator $\hat{\Theta}(x_1, \ldots, x_n)$ will be called an estimate of θ and denoted by $\hat{\theta}$.

Thus, an estimator is a random variable and an estimate is a particular realization of it.

For given $x_1, \ldots, x_n$, define

$$L(\theta) := f(x_1, \ldots, x_n; \theta).$$

It represents the likelihood that the values $x_1, \ldots, x_n$ will be observed when θ is the true value of the parameter. Thus, $L(\theta)$ is often referred to as the *likelihood function* of the random sample. Let $\hat{\theta}_{ML}$ be the maximizing value of $L(\theta)$; that is,

$$L(\hat{\theta}_{ML}) = \max_{\theta} L(\theta).$$

We refer to $\hat{\theta}_{ML}$ as the maximum likelihood estimate. Since $L(\theta)$ is a product of probability density functions (pdfs), it will always be positive (for the range of possible value of θ). Thus, $\ln L(\theta)$ can always be defined, and in determining the maximizing value of θ, it is often useful to use the fact that $L(\theta)$ and $\ln L(\theta)$ have their maximum at the same value of θ. Hence, we may also obtain $\hat{\theta}_{ML}$, by maximizing $\ln L(\theta)$.

An estimator is meant to approximate the value of the parameter. It is therefore desirable to establish how good such an approximation is. We call the difference $\hat{\Theta} - \theta$ the error of the estimator. It turns out to be a random variable.

Definition 7.1.2. The bias of an estimator $\hat{\Theta}(X_1, \ldots, X_n)$ for the parameter θ is the expectation value of the error of the estimator, i.e.

$$\mathbb{E}\hat{\Theta}(X_1, \ldots, X_n) - \theta.$$

The estimator is said to be unbiased if the bias is zero for all $\theta \in \mathcal{O}$, or in other words, the expected value of the estimator is equal to the parameter.

Example. By referring to the initial example where $f(0;\theta) = 1-\theta$ and $f(1;\theta) = \theta$, the sample mean $\sum_i x_i/n$ represents an unbiased estimator with variance $\theta(1-\theta)/n$.

The bias is the expected value of the error, and the fact that it is zero does not guarantee that the error is low with high probability. We need to look at some loss function of the error; the most commonly chosen loss function is the expected square of the error.

If $\hat{\Theta}$ is an unbiased estimator, then its mean square error is given by

$$\mathbb{E}[(\hat{\Theta}-\theta)^2] = \mathbb{E}\{[\hat{\Theta}-\mathbb{E}(\hat{\Theta})]^2\} \equiv \mathrm{Var}(\hat{\Theta}). \tag{7.2}$$

That is, its mean square error equals its variance.

Definition 7.1.3. Given two unbiased estimators $\hat{\Theta}_1$ and $\hat{\Theta}_2$ for the parameter θ, the former is said to be more efficient than the latter if $\mathrm{Var}(\hat{\Theta}_1) < \mathrm{Var}(\hat{\Theta}_2)$.

This naturally raises the question of the existence of a best estimator of θ that dominates every other. The answer to this question is provided by the Cramer–Rao lower bound on the mean square error of any estimator as we shall see.

7.1.1 *Cramer–Rao bound*

Definition 7.1.4. The score V is a random variable defined by

$$V := \frac{\partial}{\partial\theta}\ln f(X;\theta) = \frac{\frac{\partial}{\partial\theta} f(X;\theta)}{f(X;\theta)}.$$

The mean value of the score is

$$\begin{aligned}\mathbb{E}V &= \sum_x \frac{\frac{\partial}{\partial\theta} f(x;\theta)}{f(x;\theta)} f(x;\theta)\\ &= \sum_x \frac{\partial}{\partial\theta} f(x;\theta)\\ &= \frac{\partial}{\partial\theta}\sum_x f(x;\theta) = 0,\end{aligned} \tag{7.3}$$

and therefore $\mathbb{E}V^2 = \mathrm{Var}(V)$.

Definition 7.1.5. The Fisher information is defined as the variance of the score

$$J(\theta) := \mathrm{Var}(V) = \mathbb{E}\left[\frac{\partial}{\partial\theta}\ln f(x;\theta)\right]^2.$$

Its relevance appears in the following theorem.

Theorem 7.1.6 (Cramer–Rao bound). *The mean squared error of any unbiased estimator $\hat{\Theta}$ of the parameter θ is lower bounded by the reciprocal of the Fisher information:*

$$\mathrm{Var}(\hat{\Theta}) \geq \frac{1}{J(\theta)}.$$

Lemma 7.1.7. *Given two random variables X and Y, it is*

$$[\mathrm{Cov}(X,Y)]^2 \leq \mathrm{Var}(X)\,\mathrm{Var}(Y),$$

where $\mathrm{Cov}(X,Y) := \mathbb{E}(XY) - [\mathbb{E}(X)][\mathbb{E}(Y)]$.

Proof. Consider the random variable $aX + bY$ obtained by linear combination of X and Y via $a, b \in \mathbb{R}$. By the definition of variance, it must be $\mathrm{Var}(aX + bY) \geq 0$ for all $a, b \in \mathbb{R}$. This corresponds to having

$$(a,b)\begin{pmatrix} \mathrm{Var}(X) & \mathrm{Cov}(X,Y) \\ \mathrm{Cov}(X,Y) & \mathrm{Var}(Y) \end{pmatrix}\begin{pmatrix} a \\ b \end{pmatrix} \geq 0, \quad \forall a, b \in \mathbb{R}. \tag{7.4}$$

In turn, this is equivalent to requiring the semidefinite positiveness of the above matrix, which implies

$$\det\begin{pmatrix} \mathrm{Var}(X) & \mathrm{Cov}(X,Y) \\ \mathrm{Cov}(X,Y) & \mathrm{Var}(Y) \end{pmatrix} \geq 0. \tag{7.5}$$

Then the statement of the lemma follows immediately. □

Proof of Theorem 7.1.6. Applying Lemma 7.1.7 to $X = V - \mathbb{E}V$ and $Y = \hat{\Theta} - \mathbb{E}\hat{\Theta}$, we have

$$\{\mathbb{E}[(V - \mathbb{E}V)(\hat{\Theta} - \mathbb{E}\hat{\Theta})]\}^2 \leq \mathbb{E}(V - \mathbb{E}V)^2 \times \mathbb{E}(\hat{\Theta} - \mathbb{E}\hat{\Theta})^2. \tag{7.6}$$

Since $\hat{\Theta}$ is unbiased, $\mathbb{E}\hat{\Theta} = \theta$ for all θ. By (7.3), $\mathbb{E}V = 0$ and hence $\mathbb{E}(V - \mathbb{E}V) \times \mathbb{E}(\hat{\Theta} - \mathbb{E}\hat{\Theta}) = \mathbb{E}(V\hat{\Theta})$. Also, by definition, $\mathrm{Var}(V) = J(\theta)$.

Substituting these conditions in (7.6), we have

$$\{\mathbb{E}(V\hat{\Theta})\}^2 \leq J(\theta)\,\mathrm{Var}(\hat{\Theta}). \tag{7.7}$$

Now

$$\begin{aligned}
\mathbb{E}(V\hat{\Theta}) &= \sum_x \frac{\frac{\partial}{\partial\theta} f(x;\theta)}{f(x;\theta)} \hat{\Theta}(x) f(x;\theta) \\
&= \sum_x \frac{\partial}{\partial\theta} f(x;\theta)\hat{\Theta}(x) \\
&= \frac{\partial}{\partial\theta} \sum_x f(x;\theta)\hat{\Theta}(x) \\
&= \frac{\partial}{\partial\theta} \mathbb{E}\hat{\Theta} \\
&= \frac{\partial\theta}{\partial\theta} = 1.
\end{aligned}$$

Substituting this in (7.7) yields

$$\mathrm{Var}(\hat{\Theta}) \geq \frac{1}{J(\theta)},$$

which is the desired inequality. □

The Cramer–Rao inequality gives us a lower bound on the variance for all unbiased estimators. When this bound is achieved, we call the estimator *efficient*.

The Fisher information is therefore a measure of the amount of information about θ that is present in the data. It gives a lower bound on the error in estimating θ from the data. However, it is possible that there does not exist an estimator meeting this lower bound.

The Fisher information turns out to be additive. If we consider a sample of n random variables $X_1, \ldots, X_n$ drawn i.i.d., as a consequence of (7.1), we have

$$V(X_1, \ldots, X_n) = \frac{\partial}{\partial\theta} \ln f(x_1, \ldots, x_n; \theta) = \sum_{i=1}^{n} V(X_i).$$

Hence, the sample Fisher information results

$$\begin{aligned}\mathbb{E}\left[\frac{\partial}{\partial\theta}\ln f(x_1,\ldots,x_n;\theta)\right]^2 &= \mathbb{E}V(X_1,\ldots,X_n)^2\\ &= \mathbb{E}\left(\sum_{n=1}^{n}V(X_i)\right)^2\\ &= \sum_{i=1}^{n}\mathbb{E}V^2(X_i)\\ &= nJ(\theta).\end{aligned}$$

As a consequence, if a sample of n i.i.d. random variables is used in the estimation, we have the corresponding Cramer–Rao inequality

$$\mathrm{Var}(\hat{\Theta}) \geq \frac{1}{nJ(\theta)}.$$

This is consistent with the idea of reducing to zero the error in the limit of infinite size sample. Note, however, that the achievability of Cramer–Rao bound is not guaranteed. In contrast, we shall see that this will be the case in the quantum framework.

7.1.2 *Averaging the mean square error*

From a different perspective, the parameter could be considered as a random inaccessible variable Θ to be estimated through a random accessible variable X. Suppose that $\Theta \sim f(\theta)$ for some fixed (prior) probability distribution function (pdf) $f(\theta)$, then $f(x,\theta) = f(x;\theta)f(\theta)$. The estimator $\hat{\Theta}$ will be a function of the data $g(X)$.

The *average* mean square error will then be

$$\mathsf{e} := \mathbb{E}\{[g(X)-\Theta]^2\} = \sum_x \int_\theta (g(x)-\theta)^2 f(x,\theta)\,d\theta.$$

Note that the expectation here involves both random variables X and Θ, differently from Eq. (7.2), where the expectation was only with respect to X. Inverting the Bayes rule, we can further write

$$\mathsf{e} = \sum_x f(x) \left[\int_{\Theta} (g(x) - \theta)^2 f(\theta|x)\, d\theta \right].$$

This represents a cost function to be minimized. Since both f and the integrand are positive, e is minimum when the quantity inside the square brackets is minimum for all x. Thus, we impose

$$\frac{d\mathsf{e}}{dg} = 2 \int_{\Theta} (g(x) - \theta) f(\theta|x)\, d\theta = 0,$$

from which it results

$$g(x) = \int_{\Theta} \theta f(\theta|x)\, d\theta.$$

So, the best estimate (i.e. that minimizing the mean square error) is

$$g(x) = \int_{\Theta} \theta\, f(\theta|x) d\theta = \mathbb{E}(\Theta|x),$$

and consequently the best estimator is $\hat{\Theta} = \mathbb{E}(\Theta|X)$.

Note that this approach, differently from Cramer–Rao, gives a *global* optimal estimator, i.e. not depending on the value of the parameter θ (in which case the optimal estimator is considered *local*).

7.2 Quantum estimation

We shall now consider the quantum counterpart of Sections 7.1.1 and 7.1.2. In this case, we shall face a double optimization problem concerning, on the one hand, the input state and on the other hand, the output measurement process.

7.2.1 *Quantum Cramer–Rao bound*

In the quantum framework suppose we have a state ρ_θ depending on the parameter to be estimated. Then for a a POVM $\{\Pi_x\}_x$, we have $p(x|\theta) = \mathrm{Tr}[\Pi_x \rho_\theta]$.

Let us introduce the symmetric logarithmic derivative (SLD) L_θ as the Hermitian operator satisfying the equation

$$\frac{L_\theta \rho_\theta + \rho_\theta L_\theta}{2} = \frac{d\rho_\theta}{d\theta}. \tag{7.8}$$

It follows that

$$\frac{\partial}{\partial\theta} p(x|\theta) = \mathrm{Tr}\left[\frac{d\rho_\theta}{d\theta}\Pi_x\right] = \mathrm{Re}(\mathrm{Tr}[\rho_\theta \Pi_x L_\theta]).$$

The Fisher information (7.1.5) can then be written in the quantum framework as

$$J(\theta) = \sum_x \frac{\mathrm{Re}(\mathrm{Tr}[\rho_\theta \Pi_x L_\theta])^2}{\mathrm{Tr}[\rho_\theta \Pi_x]}.$$

For the given POVM $\{\Pi_x\}$, this establishes the classical bound on precision, which may be achieved by a proper data processing. On the other hand, in order to evaluate the ultimate bounds to precision, we now have to maximize the Fisher information overall quantum measurements.

$$J(\theta) \leq \sum_x \left|\frac{\mathrm{Tr}[\rho_\theta \Pi_x L_\theta]}{\sqrt{\mathrm{Tr}[\rho_\theta \Pi_x]}}\right|^2 \tag{7.9}$$

$$= \sum_x \left|\mathrm{Tr}\left[\frac{\sqrt{\rho_\theta}\sqrt{\Pi_x}}{\sqrt{\mathrm{Tr}[\rho_\theta \Pi_x]}}\sqrt{\Pi_x} L_\theta \sqrt{\rho_\theta}\right]\right|^2$$

$$\leq \sum_x \mathrm{Tr}[\Pi_x L_\theta \rho_\theta L_\theta] \tag{7.10}$$

$$= \mathrm{Tr}[L_\theta \rho_\theta L_\theta]$$

$$= \mathrm{Tr}[\rho_\theta L_\theta^2].$$

The above chain of inequalities prove the following theorem.

Theorem 7.2.1 (Quantum Cramer–Rao bound). *The Fisher information $J(\theta)$ of any quantum measurement is bounded by the* *so-called quantum Fisher information (QFI)*

$$J(\theta) \leq J_Q(\theta) := \mathrm{Tr}[\rho_\theta L_\theta^2] = \mathrm{Tr}\left[\frac{d\rho_\theta}{d\theta} L_\theta\right], \tag{7.11}$$

leading the quantum Cramer–Rao bound

$$\mathrm{Var}(\hat{\theta}) \geq \frac{1}{J_Q(\theta)}.$$

The quantum version of the Cramer–Rao theorem provides an ultimate bound: it does not depend on the measurement. Optimal quantum measurements for the estimation of θ thus correspond to POVM with a Fisher information equal to the quantum Fisher information, i.e. those saturating both inequalities (7.9) and (7.10). The first one is saturated when $\mathrm{Tr}[\rho_\theta \Pi_x L_\theta]$ is a real number $\forall\theta$. The second one is based on the Schwartz inequality

$$|\mathrm{Tr}[A^\dagger B]|^2 \leq \mathrm{Tr}[A^\dagger A]\,\mathrm{Tr}[B^\dagger B],$$

applied to $A^\dagger = \sqrt{\rho_\theta}\sqrt{\Pi_x}/\sqrt{\mathrm{Tr}[\rho_\theta \Pi_x]}$ and $B = \sqrt{\Pi_x} L_\theta \sqrt{\rho_\theta}$ and it is saturated when

$$\frac{\sqrt{\rho_\theta}\sqrt{\Pi_x}}{\sqrt{\mathrm{Tr}[\rho_\theta \Pi_x]}} = \frac{\sqrt{\Pi_x} L_\theta \sqrt{\rho_\theta}}{\mathrm{Tr}[\rho_\theta \Pi_x L_\theta]}, \quad \forall\theta. \tag{7.12}$$

The operatorial condition in Eq. (7.12) is satisfied if and only if $\{\Pi_x\}$ is made by the set of projectors over the eigenstates of L_θ, which, in turn, represents the optimal POVM to estimate the parameter θ. Note, however, that L_θ itself may not represent the optimal observable to be measured. In fact, Eq. (7.12) determines the POVM and not the estimator, i.e. the function of the eigenvalues of L_θ.

Using the fact that $\mathrm{Tr}[\rho_\theta L_\theta] = 0$, an explicit form for the optimal quantum (observable) estimator is given by

$$O_\theta = \theta I + \frac{L_\theta}{J_Q(\theta)}, \tag{7.13}$$

for which we have

$$\mathrm{Tr}[O_\theta \rho_\theta] = \theta,$$

$$\mathrm{Tr}[O_\theta^2 \rho_\theta] = \theta^2 + \frac{\mathrm{Tr}[\rho_\theta L_\theta^2]}{J_Q^2(\theta)},$$

and thus $\mathbb{E}[O_\theta - \mathbb{E}O_\theta]^2 = 1/J_Q(\theta)$. This shows that the quantum Cramer–Rao bound, differently from the classical one, can always be attained.

We now prove that also the QFI is additive, that is

$$\rho = \sigma \otimes \tau \Rightarrow J_{Q,\rho}(\theta) = J_{Q,\sigma}(\theta) + J_{Q,\tau}(\theta). \tag{7.14}$$

As $\frac{d}{d\theta}\rho = \frac{d}{d\theta}\sigma \otimes \tau + \sigma \otimes \frac{d}{d\theta}\tau$ it is

$$\begin{aligned}\mathrm{Tr}\left[\frac{d\rho_\theta}{d\theta}L_\theta\frac{d\rho_\theta}{d\theta}\right] = \mathrm{Tr}\Bigg[&\frac{d\sigma_\theta}{d\theta}L_\theta\frac{d\sigma_\theta}{d\theta}\otimes\tau + \frac{d\sigma_\theta}{d\theta}\otimes\tau L_\theta\frac{d\tau_\theta}{d\theta}\\ &+\sigma\otimes\frac{d\tau_\theta}{d\theta}L_\theta\frac{d\tau_\theta}{d\theta} + \sigma L_\theta\frac{d\sigma_\theta}{d\theta}\otimes\frac{d\tau_\theta}{d\theta}\Bigg],\end{aligned}$$

and the results follow by noting that $\mathrm{Tr}(\frac{d\sigma_\theta}{d\theta}) = \mathrm{Tr}(\frac{d\tau_\theta}{d\theta}) = 0$.

Therefore, if n replicas of ρ, namely $\rho^{\otimes n}$, are used in the estimation, we have the corresponding Cramer–Rao inequality

$$\mathrm{Var}(\theta) \geq \frac{1}{nJ_Q(\theta)}. \tag{7.15}$$

7.2.2 *Averaging the quantum mean square error*

Suppose we have a random variable Θ that takes on values θ with probability density $p(\theta)$. This random variable is coupled to the state of a quantum system through the conditional density operator ρ_θ. We wish to perform a measurement by Hermitian operator $\hat{\Theta}$, the outcome of which is an estimate of the value θ. We wish to minimize the mean square error between the measurement outcome and the value θ assumed by the random variable Θ, which is

$$C(\hat{\Theta}) = \int_{\mathcal{O}} p(\theta)\mathrm{Tr}[\rho_\theta(\hat{\Theta} - \theta I)^2]d\theta. \tag{7.16}$$

The approach is to first find the Hermitian operator that minimizes $C(\hat{\Theta})$. We define this operator as the minimum mean square error (MMSE) estimator. We find the measurement associated with the MMSE estimator after we find the estimator (operator).

Theorem 7.2.2. *Defining the following operators*

$$W^{(0)} := \int_{\mathcal{O}} p(\theta)\rho_\theta d\theta, \quad W^{(1)} := \int_{\mathcal{O}} \theta p(\theta)\rho_\theta d\theta \tag{7.17}$$

the MMSE estimator is obtained as a solution of the following linear (matrix) equation

$$W^{(0)}\hat{\Theta}_{\text{opt}} + \hat{\Theta}_{\text{opt}}W^{(0)} = 2W^{(1)}. \tag{7.18}$$

Proof. Let ϵ be a real number and H any Hermitian operator. Let $\hat{\Theta}_{\text{opt}}$ be the MMSE estimator. We must have $C(\hat{\Theta}_{\text{opt}}+\epsilon H) \geq C(\hat{\Theta}_{\text{opt}})$ since the sum of two Hermitian operators is an Hermitian operator. Writing this out, we obtain

$$\begin{aligned} C(\hat{\Theta}_{\text{opt}}) \leq C(\hat{\Theta}_{\text{opt}}) &+ \epsilon \text{Tr}[H(W^{(0)}\hat{\Theta}_{\text{opt}} + \hat{\Theta}_{\text{opt}}W^{(0)} - 2W^{(1)})] \\ &+ \epsilon^2 \text{Tr}[W^{(0)}H^2], \end{aligned}$$

for all ϵ and Hermitian H. Differentiating with respect to ϵ, we obtain the necessary condition

$$\text{Tr}[H(W^{(0)}\hat{\Theta}_{\text{opt}} + \hat{\Theta}_{\text{opt}}W^{(0)} - 2W^{(1)})] = 0, \tag{7.19}$$

for all Hermitian H. We next show that a Hermitian solution to (7.19) must satisfy (7.18). Define G as

$$G := W^{(0)}\hat{\Theta}_{\text{opt}} + \hat{\Theta}_{\text{opt}}W^{(0)} - 2W^{(1)}.$$

Since $W^{(0)}$ and $W^{(1)}$ are Hermitian, and since $\hat{\Theta}$ is Hermitian, G must be Hermitian. Let $|g_i\rangle$ be an eigenvector of G with eigenvalue g_i. If we let H equal the Hermitian operator $|g_i\rangle\langle g_i|$, we have

$$\text{Tr}[G|g_i\rangle\langle g_i|] = \langle g_i|G|g_i\rangle = g_i = 0.$$

Thus, all the eigenvalues of G are zero, and G is the null operator. This means that a Hermitian solution to (7.19) must satisfy (7.18). □

If $W^{(0)}$ is positive definite, it is possible to show that the solution of (7.18) is unique.

Let us now study the POVM that allows us to realize the MMSE. It will be constructed by means of the eigenstates of the minimizing operators $\hat{\Theta}$. Let us write the spectral decomposition of $\hat{\Theta}$ as

$$\hat{\Theta} = \sum_i \hat{\theta}_i |\hat{\theta}_i\rangle\langle\hat{\theta}_i| \equiv \int_{\mathcal{O}} \hat{\theta}\, \Pi(\hat{\theta})\, d\hat{\theta}, \tag{7.20}$$

where

$$\Pi(\hat{\theta}) := \sum_i \delta(\hat{\theta} - \hat{\theta}_i)|\hat{\theta}_i\rangle\langle\hat{\theta}_i|. \tag{7.21}$$

Here, avoiding technical details, the quantity $\delta(\hat{\theta} - \hat{\theta}_i)$ is intended to act upon integration like a Kronecker delta acts upon summation.

Inserting (7.20) into (7.16), we can find

$$\begin{aligned} C(\hat{\Theta}) &= \int_{\mathcal{O}}\int_{\mathcal{O}} (\hat{\theta} - \theta)^2 \mathrm{Tr}[\Pi(\hat{\theta})\rho_\theta] p(\theta)\, d\hat{\theta}\, d\theta, \\ &= \int_{\mathcal{O}}\int_{\mathcal{O}} (\hat{\theta} - \theta)^2 p(\hat{\theta}|\theta) p(\theta)\, d\hat{\theta}\, d\theta. \end{aligned} \tag{7.22}$$

Note that the latter represents an explicit expression of the average quadratic cost function, that is the mean square error averaged over all possible values of the parameter θ. Thus, once Eq. (7.18) has been solved, the eigenvalues $\hat{\theta}_i$ of $\hat{\Theta}$ can be found and by the POVM (7.21), the minimum of (7.22) evaluated.

7.3 Entanglement-assisted quantum estimation

We show in this section how entanglement can be helpful in some circumstances to enhance the estimation capabilities of a quantum channel. We distinguish between *locally* optimal strategy and *globally* optimal strategy.

7.3.1 *Beating the quantum Cramer–Rao bound*

Consider the depolarizing quantum channel

$$\mathcal{N}_\theta(\rho) = \theta\rho + \frac{1-\theta}{2}I, \tag{7.23}$$

where $\theta \in [-\frac{1}{3}, 1]$ (see Section 6.5).

Our task is to find an optimal input for the channel that maximizes the QFI of the corresponding parametric family of output states.

Since the depolarizing channel is isotropic, we can take without loss of generality the optimal input to be $\rho = |0\rangle\langle 0|$. The corresponding output state will be

$$\rho_\theta = \mathcal{N}_\theta(\rho) = \frac{1}{2}(1+\theta)|0\rangle\langle 0| + \frac{1}{2}(1-\theta)|1\rangle\langle 1|.$$

Simple calculations according to (7.8) and (7.11) give the Fisher information

$$J_Q = \frac{1}{1-\theta^2}.$$

On two identical uses of the channel with the same input state, J_θ will be doubled due to its additivity, i.e. $J_Q = 2/(1-\theta^2)$.

We next study the extended channel $\mathcal{N}_\theta \otimes \mathrm{id}$. In this case, we can use a possibly entangled state as the input. By the same reason as above, we can take the input to be a pure state: $\rho = |\psi\rangle\langle\psi|$. By the Schmidt decomposition, the vector is represented as

$$|\psi\rangle = \sqrt{x}|e_0\rangle|f_0\rangle + \sqrt{1-x}|e_1\rangle|f_1\rangle, \tag{7.24}$$

where $x \in [0,1]$ and $\{|e_0\rangle, |e_1\rangle\}$ and $\{|f_0\rangle, |f_1\rangle\}$ are orthonormal bases of $\mathbb{C}^2$. The corresponding output state $\hat{\rho}_\theta := \mathcal{N}_\theta \otimes \mathrm{id}(\sigma)$ becomes

$$\hat{\rho}_\theta = \frac{1}{2}\begin{pmatrix} (1-x)(1+\theta) & 0 & 0 & 2\sqrt{x(1-x)}\theta \\ 0 & x(1-\theta) & 0 & 0 \\ 0 & 0 & (1-x)(1-\theta) & 0 \\ 2\sqrt{x(1-x)}\theta & 0 & 0 & x(1+\theta) \end{pmatrix}.$$

The Fisher information, again by means of (7.8) and (7.11), results in this case

$$\hat{J}_Q = \frac{1 + 3\theta + 8x(1-x)}{(1-\theta^2)(1+3\theta)}.$$

It takes the maximum $\hat{J}_Q = 3/(1-\theta)(1+3\theta)$ at $x = 1/2$. Therefore, the optimal input for the channel $\mathcal{N}_\theta \otimes \mathrm{id}$ is the maximally entangled state.

Last, let us proceed to the analysis of the channel $\mathcal{N}_\theta \otimes \mathcal{N}_\theta$ with the same input as in (7.24). Here, the output reads

$$\check{\rho}_\theta = \frac{1}{4}\begin{pmatrix} -4x\theta + (1+\theta)^2 & 0 & 0 & 4\sqrt{x(1-x)}\theta^2 \\ 0 & 1-\theta^2 & 0 & 0 \\ 0 & 0 & 1-\theta^2 & 0 \\ 4\sqrt{x(1-x)}\theta^2 & 0 & 0 & 4x\theta + (1-\theta)^2 \end{pmatrix}.$$

The Fisher information, using (7.8) and (7.11), turns out to be

$$\check{J}_\theta = \frac{4\theta^4 + 5\theta^2 - 1 + 16\theta^4 x(1-x)}{2\theta^2(1-\theta^4)} + \frac{1-\theta^2}{2\theta^2(1+\theta^2)(1-\theta^2+16\theta^2 x(1-x))}.$$

For $-1/3 \le \theta \le 1/\sqrt{3}$, the Fisher information $\check{J}_\theta$ takes the maximum $2/(1-\theta^2)$ at $x = 0$ and 1, while for $1/\sqrt{3} \le \theta \le 1$ it takes the maximum $12\theta^2/(1-\theta^2)(1+3\theta^2)$ at $x = \frac{1}{2}$.

By comparison of the three cases, the best strategy for estimating the isotropic depolarization parameter θ is determined to be the following:

- For $-\frac{1}{3} \le \theta \le \frac{1}{3}$, use $\mathcal{N}_\theta \otimes \mathrm{id}$ and input a maximally entangled state;
- For $\frac{1}{3} \le \theta \le \frac{1}{\sqrt{3}}$, use $\mathcal{N}_\theta \otimes \mathcal{N}_\theta$ and input any pure factorable state;
- For $\frac{1}{\sqrt{3}} \le \theta \le 1$, use $\mathcal{N}_\theta \otimes \mathcal{N}_\theta$ and input a maximally entangled state.

Not only does this show the possibility of beating the Cramer–Rao bound for separable states by employing entanglement, but it is also a manifestation of the locality behavior, because the optimal strategy changes according to the value of θ.

7.3.2 *Reducing the average quantum mean square error*

Let us now consider the goal of enhancing the performance of global estimation by means of entanglement. Consider two possibilities,

$$\Psi_1(\theta) = (\mathcal{N}_\theta \otimes \mathrm{id})|\psi\rangle\langle\psi|,$$
$$\Psi_2(\theta) = (\mathcal{N}_\theta \otimes \mathcal{N}_\theta)|\psi\rangle\langle\psi|,$$

where the input state vector reads like the previous section (Eq. (7.24)),

$$|\psi\rangle = \sqrt{x}|e_0\rangle|f_0\rangle + \sqrt{1-x}|e_1\rangle|f_1\rangle.$$

Then, according to (7.22), the average for the quadratic cost function is given by

$$C_i(x) = \int_{-1/3}^{1}\int_{-1/3}^{1} (\hat{\theta}-\theta)^2 p(\theta)\,\mathrm{Tr}[\Pi(\hat{\theta})\Psi_i(\theta)]d\hat{\theta}\,d\theta,$$

where, assuming to have no *a priori* knowledge about θ, we take a flat distribution $p(\theta) = \frac{3}{4}$.

In order to apply Theorem 7.2.2, let the spectral decomposition of $W^{(0)}$ be

$$W^{(0)} = \sum_{i=1}^{4} \omega_i |\omega_i\rangle\langle\omega_i|.$$

Then the minimizing operator $\hat{\Theta}$ results

$$\hat{\Theta} = \sum_{i,j=1}^{4} \frac{2}{\omega_i+\omega_j} |\omega_i\rangle\langle\omega_i|W^{(1)}|\omega_j\rangle\langle\omega_j|.$$

From this, one can find the eigenvalues $\hat{\theta}_i$ and using (7.21) arrive at minimum of the average quadratic cost function.

For Ψ_1 it reads

$$C_1(x) = \frac{8}{81}\left[1 + \left(x - \frac{1}{2}\right)^2\right],$$

which is minimized by maximally entangled state ($x = 1/2$).

Instead, for Ψ_2, it is

$$C_2(x) = \frac{8}{2295}\frac{391 + 606x(1-x) - 10x^2(1-x)^2}{13 + 8x(1-x)},$$

whose minimum is attained at $x = 0, 1$, by separable states.

We conclude that $\min C_1 = \frac{8}{81} < \min C_2 = \frac{184}{1755}$ so that the optimal estimation strategy, using two probe qubits, is to prepare them as a maximally entangled pair and to input one qubit of the pair into the channel keeping the other untouched. In this case, the estimation is obtained by applying the two POVM elements, $\Pi_1 = |\psi\rangle\langle\psi|$ and $\Pi_2 = I - |\psi\rangle\langle\psi|$, corresponding to $\hat{\theta}_1 = 5/9$ and $\hat{\theta}_2 = \hat{\theta}_3 = \hat{\theta}_4 = 1/9$.

We close this section saying that unfortunately there is no general rule for usefulness of entanglement. This depends on the kind of channel as well as on its dimension.

Exercises

1. Let $X_1, \ldots X_n$ be a random sample of X with mean μ and variance σ^2. Show that $\frac{1}{n}\sum_{i=1}^n X_i$ is the most efficient linear unbiased estimator of μ (where linearity is referred to the arguments $X_1, \ldots, X_n$).
2. Let $X_1, \ldots, X_n$ be a random sample of a binomial random variable X with parameters (m, p), where m is assumed to be known and p unknown. Determine the maximum-likelihood estimator of p.
3. Prove the convexity of the QFI, i.e. if $\rho_\theta = \lambda\sigma_\theta + (1-\lambda)\tau_\theta$ for a constant λ between 0 and 1, then $J_Q(\rho_\theta) \le \lambda J_Q(\sigma_\theta) + (1-\lambda)$ $J_Q(\tau_\theta)$. This property allows us to just focus on pure states for maximizing the QFI.
4. Find the quantum Cramer–Rao bound for the unitary channel $\mathcal{U}_\theta\rho = e^{-i\theta G}\rho e^{i\theta G}$ with G Hermitian.

5. Repeat the approach of Section 7.3.2 to the case of d-dimensional depolarizing quantum channel

$$\mathcal{N}_\theta(\rho) = \theta\rho + \frac{1-\theta}{d}I, \quad -\frac{1}{d^2-1} \leq \theta \leq 1,$$

comparing maximally entangled input states with separable ones.

Chapter 8

DATA COMPRESSION

In this Chapter, we introduce the notions of classical and quantum sources that consist of to ensembles of states. Then, the compression of data emitted by such sources will be discussed and its limit established. This provides an operational meaning to both Shannon and von Neumann entropy.

8.1 Classical data compression

A source that emits symbols x belonging to an alphabet $\mathcal{X}$ according to a probability distribution $p(x)$ can be identified with a random variable X.

Let us consider a sequence of n random variables, $X_1, X_2, \ldots, X_n$. They are said to be independent and identically distributed (i.i.d.) if they are statically independent and they all have the same alphabet $\mathcal{X}$ and the same probability distribution p.

Recall that a sequence of n random variables $X_1, X_2, \ldots, X_n$ is i.i.d. if they are n identical and independent copies of the same random variables, denoted X. We can say that the sequence of i.i.d. variables is drawn from the distribution p_X.

A sequence of n i.i.d. random variables produces a sequence

$$x_1 x_2 \cdots x_n, \quad x_i \in \mathcal{X}, \tag{8.1}$$

with probability $p(x_1 x_2 \cdots x_n) = \prod_{i=1}^{n} p_X(x_i)$. Since $|\mathcal{X}|$ is the number of symbols in the alphabet, there will be $|\mathcal{X}|^n$ of such sequences. So, in principle the number of bits necessary to represent (or store) this source would be $\log |\mathcal{X}|^n$. Can we reduce it, i.e. compress the source?

Example. Let us consider a random variable X, assuming four different values:

$$\mathcal{X} = \{A, B, C, D\},$$

with probabilities $p(A) = 1/2$, $p(B) = 1/4$, $p(C) = p(D) = 1/8$.

We can define a map from the alphabet $\mathcal{X}$ to binary strings in the following way:

$$A \to 0, \quad B \to 10, \quad C \to 110, \quad D \to 111.$$

Then, the expected length of strings results as $\frac{1}{2} \times 1 + \frac{1}{4} \times 2 + \frac{1}{8} \times 3 + \frac{1}{8} \times 3 = \frac{7}{4} < 2$. What we have done is to store the most (resp., least) frequent symbols in the shortest (resp., longest) strings.

This shows that by accounting for the occurrence probability of symbols, we can compress the source.

Let us consider an encoding map $E_n : \mathcal{X}^n \to \mathbb{F}_2^k$, mapping each of the possible sequences (8.1) into binary strings of smaller length called codewords[1]

$$x_1 x_2 \ldots x_n \to E_n(x_1 x_2 \cdots x_n). \tag{8.2}$$

Generally, this encoding is not invertible (it will simply be a stochastic map or classical channel). It is hence required to define a decoding map $\Delta_n : \mathbb{F}_2^k \to \mathcal{X}^n$, such that

$$E_n(x_1 x_2 \ldots x_n) \to \Delta_n(E_n(x_1 x_2 \cdots x_n)) = x_1' x_2' \cdots x_n'. \tag{8.3}$$

Since the encoding map is not invertible, it may be possible that $x_1' x_2' \cdots x_n' \neq x_1 x_2 \cdots x_n$, which leads to an error in the compression scheme.

[1]In contrast to the above example, from now on we restrict our attention to fixed length codewords.

The compression scheme with encoding map E_n and decoding map Δ_n is characterized by a *compression rate* $\frac{k}{n} \in (0, \log|\mathcal{X}|]$ and a probability of error

$$p_e = \Pr\{\Delta_n(E_n(X_1X_2 \ldots X_n)) \neq X_1X_2 \cdots X_n\}. \tag{8.4}$$

Clearly, a "good" compression scheme should have small probability of error, at least in the limit $n \to \infty$. Thus, the notion of compression scheme concerns a sequence of encoding and decoding maps $\{E_n, \Delta_n\}$. It is then clear that over such a sequence, k depends on n.

Definition 8.1.1. A compression rate $\mathtt{R}$ is said to be achievable if there exists a sequence of encoding/decoding maps $\{E_n, \Delta_n\}_n$ such that

$$\lim_{n\to\infty} \frac{k}{n} = \mathtt{R}, \quad \text{and} \quad \lim_{n\to\infty} p_e = 0.$$

The ultimate limit for compression schemes is stated by the source coding (or noiseless channel coding) theorem by Shannon.

8.1.1 *Shannon's noiseless coding theorem*

Theorem 8.1.2. *Given an i.i.d. source X on $\mathcal{X}$, any $\mathtt{R} \geq H(X)$ is an achievable compression rate, while no $\mathtt{R} < H(X)$ is.*

The notion of *typical sequences* is needed to prove this theorem.

Definition 8.1.3 (Typical sequences). For any block length n and any $\epsilon > 0$, the set of typical sequences $T_X(n, \epsilon)$ is defined as the set of sequences $x_1x_2 \ldots x_n$ such that

$$2^{-n(H(X)+\epsilon)} \leq p(x_1x_2 \cdots x_n) \leq 2^{-n(H(X)-\epsilon)}. \tag{8.5}$$

This definition agrees with our intuition. In fact, for an i.i.d. source that emits the symbol $x \in \mathcal{X}$ with probability $p(x)$, we expect a sequence of length n to contain, with high probability, $\sim np(x)$

repetitions of the symbol x. Then the probability of such a sequence results

$$\approx \prod_{x\in\mathcal{X}} p(x)^{np(x)} = \prod_{x\in\mathcal{X}} 2^{\log p(x)^{np(x)}} = 2^{n\sum_{x\in\mathcal{X}} p(x)\log p(x)}$$

$$= 2^{-nH(X)}. \tag{8.6}$$

The typical sequences have the following properties:

(i) The probability that a sequence $x_1, x_2, \ldots, x_n$ randomly picked from $\mathcal{X}^n$ belongs to the set of typical sequences tends to 1 in the limit $n \to \infty$. That is, for any $\epsilon > 0$, there exists an $n_\epsilon \in \mathbb{N}$ such that

$$\Pr\{X^n \in T_X(n,\epsilon)\} \geq 1 - \epsilon,$$

for all $n \geq n_\epsilon$.
This property directly follows from the *law of large numbers*. Given a sequence of i.i.d. real random variables $L, L_1, L_2, \ldots, L_n$, with expectation values $\mathbb{E}(L)$, $\mathbb{E}(L^2) < +\infty$, and all $\epsilon > 0$,

$$\lim_{n\to\infty} \Pr\left\{\left|\frac{1}{n}\sum_{i=1}^{n} L_i - \mathbb{E}(L)\right| > \epsilon\right\} = 0.$$

(ii) For any $\epsilon > 0$ and for sufficiently large n,

$$(1-\epsilon)2^{n(H(X)-\epsilon)} \leq |T_X(n,\epsilon)| \leq 2^{n(H(X)+\epsilon)}.$$

This property directly follows from the definition of typicality and the observation that the sum of probabilities of the typical sequences must lie in $[1-\epsilon, 1]$.

(iii) Given any set $S(n)$ of sequences containing at most $2^{n\mathtt{R}}$ elements, with $\mathtt{R} < H(X)$, the probability that a random sequence belongs to the set $S(n)$ goes to zero by $n \to \infty$.

To show this property, we have to split $S(n)$ into two subsets: the one of typical sequences and the one of non-typical sequences. Then, from property (i) we already know that the probability of having non-typical sequences in $S(n) \subseteq \mathcal{X}^n$ goes to zero for $n \to \infty$. For what concerns the typical sequences, its number in $S(n)$ will be smaller than $|S(n)| \leq 2^{n\mathtt{R}}$. Since each typical sequence has probability $\leq 2^{-nH(X)+n\epsilon}$, it follows that the total probability of typical sequences in $S(n)$ goes like $2^{n(\mathtt{R}-H(X)+\epsilon)}$, tending to zero for $n \to \infty$ when $\mathtt{R} + \epsilon < H(X)$.

Proof of Theorem 8.1.2. Suppose first $\mathtt{R} > H(X)$. Choose $\epsilon > 0$ such that $H(X)+\epsilon \leq \mathtt{R}$. Consider $T(n,\epsilon)$, then we know by property (ii) that, for sufficiently large n, there are at most $2^{n(H(X)+\epsilon)} \leq 2^{nR}$ typical sequences. The compression scheme then works as follows:

- Divide all sequences in $\mathcal{X}^n$ into two sets: the typical set $T(n,\epsilon)$ and the atypical set $A(n,\epsilon)$.
- Order all elements in $T(n,\epsilon)$ according to some order (e.g. lexicographic). Then we can represent each sequence in $T(n,\epsilon)$ by giving the index of the sequence in the set. Since

$$|T(n,\epsilon)| \leq 2^{n(H(X)+\epsilon)} \leq 2^{n\mathtt{R}},$$

  the indexing requires no more than $\lceil n\mathtt{R} \rceil$ bits.
- Examine each output of the source to see if it is typical. If so, we compress it by simply storing its index. The original sequence can be later recovered unambiguously from the corresponding $n\mathtt{R}$ bit string (by property (i)).
- If the output string is not typical, then we can compress it to some dummy string of $n\mathtt{R}$ bit, e.g. $(00\ldots0)$.
- In this latter case, the decompression scheme fails. However, these cases have a vanishingly small probability for $n \to \infty$ by property (i).

The case $\mathtt{R} = H(X)$ implies $\epsilon = 0$, hence it can only be achieved when $n \to \infty$ according to properties of typical sequences.

Now suppose $\mathtt{R} < H(X)$. From property (iii) of typical sequences, we know that for any set $S(n)$ of sequences containing at most $2^{n\mathtt{R}}$ elements with $\mathtt{R} < H(X)$, the probability that a random sequence belongs to $S(n)$ goes to zero. Hence, it will be so for typical sequences too. □

8.2 Quantum data compression

A quantum source that emits state vectors $|\psi_x\rangle \in \mathcal{H}$ labeled by a symbol $x \in \mathcal{X}$ and according to a probability distribution $p(x)$ defines a density operator $\rho = \sum_{x\in\mathcal{X}} p(x)|\psi_x\rangle\langle\psi_x|$.

We now consider n i.i.d. realizations of the source, yielding the product state $\rho^{\otimes n}$ defined on the Hilbert space $\mathcal{H}^{\otimes n}$. To compress this quantum source, we take an encoding CPTP map

$$\mathcal{E}_n : \mathfrak{D}(\mathcal{H}^{\otimes n}) \to \mathfrak{D}((\mathbb{C}^2)^{\otimes k}).$$

Furthermore, we need a decompression CPTP map

$$\mathcal{D}_n : \mathfrak{D}((\mathbb{C}^2)^{\otimes k}) \to \mathfrak{D}(\mathcal{H}^{\otimes n}). \tag{8.7}$$

Like in the classical case of Section 8.1, the compression scheme given by the CPTP maps $\mathcal{E}_n$, $\mathcal{D}_n$ will be characterized by a compression rate $\frac{k}{n} \in (0, \log|\mathcal{H}|]$ and a probability of error

$$p_e = 1 - F(\rho^{\otimes n}, \mathcal{D}_n \circ \mathcal{E}_n), \tag{8.8}$$

where F is the entanglement fidelity (defined in Eq. (6.12)).

Definition 8.2.1. A compression rate $\mathtt{R}$ is said to be achievable if there exists a sequence of encoding/decoding maps $\{\mathcal{E}_n, \mathcal{D}_n\}_n$ such that

$$\lim_{n\to\infty} \frac{k}{n} = \mathtt{R}, \quad \text{and} \quad \lim_{n\to\infty} p_e = 0.$$

Theorem 8.2.2 (Schumacher theorem). *Given an i.i.d. source (of pure states) described by the density operator ρ on $\mathcal{H}$, any $\mathtt{R} \geq S(\rho)$ is an achievable compression rate, while no $\mathtt{R} < S(\rho)$ is.*

The notion of *typical subspaces* is needed to prove this theorem.

Consider the spectral decomposition $\rho = \sum_i \lambda_i |i\rangle\langle i|$. Then,

$$\rho^{\otimes n} = \sum_{i_1,\ldots,i_n} \lambda_{i_1}\lambda_{i_2}\cdots\lambda_{i_n} |i_1\rangle\langle i_1| \otimes |i_2\rangle\langle i_2| \otimes \cdots \otimes |i_n\rangle\langle i_n|. \tag{8.9}$$

Clearly, the probability $\lambda_{i_1}\lambda_{i_2}\cdots\lambda_{i_n}$ can be understood as that associated to n i.i.d. copies of the classical random variable X.

Definition 8.2.3 (Typical subspaces). The typical subspace $\mathcal{T}_\rho(n,\epsilon)$ is that spanned by the typical sequences, i.e.

$$\mathcal{T}_\rho(n,\epsilon) := \text{span}\{|i_1, i_2, \cdots, i_n\rangle : i_1 i_2 \cdots i_n \text{ typical}\}.$$

The projection operator onto the typical subspace is hence given by

$$\Pi_\rho(n,\epsilon) = \sum_{i_1 i_2 \cdots i_n \in T_X(n,\epsilon)} |i_1\rangle\langle i_1| \otimes |i_2\rangle\langle i_2| \cdots \otimes |i_n\rangle\langle i_n|.$$

The typical subspaces have the following properties that straightforwardly follow from the properties of the typical sequences:

(i) For any ϵ, and for sufficiently large n,

$$\text{Tr}[\Pi_\rho(n,\epsilon)\rho^{\otimes n}] \geq 1 - \epsilon.$$

(ii) For any ϵ, and for sufficiently large n,

$$(1-\epsilon)2^{n(S(\rho)-\epsilon)} \leq |\mathcal{T}_\rho(n,\epsilon)| \leq 2^{n(S(\rho)+\epsilon)}.$$

(iii) Let $\Sigma(n)$ be a projection operator onto any subspace of $(\mathbb{C}^2)^{\otimes n}$ of dimension not bigger than $2^{n\mathtt{R}}$ with $\mathtt{R} < S(\rho)$. Then for any $\epsilon > 0$, and for sufficiently large n,

$$\text{Tr}[\Sigma(n)\rho^{\otimes n}] \leq \epsilon.$$

Proof of the Schumacher Theorem 8.2.2. Take $\mathtt{R} > S(\rho)$ and $\epsilon > 0$ such that $\mathtt{R} \geq S(\rho) + \epsilon$. By considering the typical subspace, we get $\text{Tr}[\Pi_\rho(n,\epsilon)\rho^{\otimes n}] \geq 1 - \epsilon$ (by property (i)) and $|\mathcal{T}_\rho(n,\epsilon)| \leq 2^{n\mathtt{R}}$ (by property (ii)).

Let us then consider a space $(\mathbb{C}^2)^{\otimes \lceil n\mathtt{R} \rceil}$ such that we can identify $\mathcal{T}(n,\epsilon)$ with a subspace of $(\mathbb{C}^2)^{\otimes \lceil n\mathtt{R} \rceil}$. To define a reliable compression scheme, we proceed as follows. First, make a measurement to decide if the n-system quantum state belongs to the typical subspace. To do that, one has to perform the POVM with elements $\Pi_\rho(n,\epsilon)$ (outcome 0: state belongs to typical subset) and $I - \Pi_\rho(n,\epsilon)$ (outcome 1: state belongs to the non-typical subspace, that is, the orthogonal complement of the typical subspace). If the outcome 0 is obtained, one does nothing. On the other hand, if outcome 1 is obtained, one replaces the state with some "dummy" vector, say $|\emptyset\rangle \in (\mathbb{C}^2)^{\otimes\lceil n\mathtt{R}\rceil}$. Thus, the compressing map results

$$\mathcal{E}_n(\sigma) = \Pi_\rho(n,\epsilon)\sigma\Pi_\rho(n,\epsilon) + \sum_i A_i\sigma A_i^\dagger,$$

where $A_i = |\emptyset\rangle\langle i|$ with $\{|i\rangle\}_i$ a basis for the non-typical subspace.

We can then define the decompressing map $\mathcal{D}_n$ as the identity CPTP map on $(\mathbb{C}^2)^{\otimes\lceil n\mathtt{R}\rceil}$. The entanglement fidelity of the compressing scheme thus results

$$\begin{aligned} F(\rho^{\otimes n}, \mathcal{D}_n \circ \mathcal{E}_n) &= |\mathrm{Tr}[\rho^{\otimes n}\Pi(n,\epsilon)]|^2 + \sum_i |\mathrm{Tr}(\rho^{\otimes n}A_i)|^2 \\ &\geq |\mathrm{Tr}[\rho^{\otimes n}\Pi(n,\epsilon)]|^2 \\ &\geq |1-\epsilon|^2. \end{aligned}$$

The case $\mathtt{R} = S(\rho)$ implies $\epsilon = 0$, hence it can only be achieved when $n \to \infty$ according to the property of typical subspaces.

To prove the converse of the theorem, let us assume that there exists a compression map $\mathcal{E}_n$ (with Kraus operators E_j) from $\mathfrak{D}(\mathcal{H}^{\otimes n})$ to $\mathfrak{D}((\mathbb{C}^2)^{\lceil n\mathtt{R}\rceil})$ with $\mathtt{R} < S(\rho)$. Further, suppose that the corresponding decoding map $\mathcal{D}_n$ has Kraus operators D_k. Then, the entanglement fidelity for such a compressing scheme reads

$$F(\rho^{\otimes n}, \mathcal{D}_n \circ \mathcal{E}_n) = \sum_{j,k} |\mathrm{Tr}(D_k E_j \rho^{\otimes n})|^2. \tag{8.10}$$

Let us denote the projector onto the range of D_k by $\Sigma^k(n)$. As the domain of D_k is the space $(\mathbb{C}^2)^{\lceil n\mathtt{R}\rceil}$, its range is a subspace of $\mathcal{H}^{\otimes n}$ of dimension not bigger than $2^{\lceil n\mathtt{R}\rceil}$. Hence, $D_k = \Sigma^k(n)D_k$. Then we have

$$\begin{aligned} F(\rho^{\otimes n}, \mathcal{D}_n \circ \mathcal{E}_n) &= \sum_{j,k} |\mathrm{Tr}[\Sigma^k(n) D_k E_j \rho^{\otimes n}]|^2 \\ &= \sum_{j,k} |\mathrm{Tr}[D_k E_j \rho^{\otimes n}\Sigma^k(n)]|^2 \end{aligned}$$

$$
\begin{aligned}
&= \sum_{j,k} |\mathrm{Tr}[D_k E_j \sqrt{\rho^{\otimes n}}\, \sqrt{\rho^{\otimes n}} \Sigma^k(n)]|^2 \\
&\leq \sum_{j,k} \mathrm{Tr}(D_k E_j \rho^{\otimes n} E_j^\dagger D_k^\dagger)\, \mathrm{Tr}[\Sigma^k(n) \rho^{\otimes n}] \\
&\leq \epsilon \sum_{j,k} \mathrm{Tr}(D_k E_j \rho^{\otimes n} E_j^\dagger D_k^\dagger) \\
&\leq \epsilon,
\end{aligned}
$$

where the inequality on the fourth line comes from considering the Schwarz inequality in the space $\mathfrak{L}(\mathcal{H}^{\otimes n})$, i.e. $|\mathrm{Tr}(A^\dagger B)|^2 \leq \mathrm{Tr}(A^\dagger A)\,\mathrm{Tr}(A^\dagger A)$, for any $A, B \in \mathfrak{L}(\mathcal{H}^{\otimes n})$. □

Exercises

1. Suppose we want to transmit pictures consisting of rectangular arrays of t black-and-white pixels through a transmission line that accepts a single bit at each use. If we identify the black (resp., white) pixels with 0 (resp., 1), the required time is t seconds. If we know that 83% of the spots are black and 17% are white, show that by a proper coding of the spots the transmission time can be reduced by a factor of 0.65.
2. Given a classical source, an optimal code, in the sense of shortest expected length, can be constructed by the following algorithm (Huffman coding). First the probabilities of the various symbols are arranged in decreasing order. Then, for a binary code, the two longest codewords (assigned to the two lowest probable symbols) must differ only in the last bit. The probabilities of these two symbols are hence merged and the new probabilities are again arranged in decreasing order. Proceeding in this way, the probability vector is shortened at each step. The procedure ends up when such a vector reduces to a single value 1.

   Now consider a classical source X which emits four symbols with probabilities $(\frac{1}{3}, \frac{1}{3}, \frac{1}{5}, \frac{2}{15})$. Construct a Huffman code for this source, showing that there exist two different sets of optimal lengths for the codewords.

3. Show that the trace distance between $\rho^{\otimes n}$ and $\Pi(n,\epsilon)\rho^{\otimes n}\Pi(n,\epsilon)$, i.e. $\|\rho^{\otimes n} - \Pi(n,\epsilon)\rho^{\otimes n}\Pi(n,\epsilon)\|_1$, is smaller than $2\sqrt{\epsilon}$, once the probability of having $\rho^{\otimes n}$ into the typical subspace is $\mathrm{Tr}\{\Pi(n,\epsilon)\rho^{\otimes n}\} \geq 1-\epsilon$.
4. Suppose a quantum source generates two non-orthogonal state vectors $|\phi_+\rangle$ and $|\phi_-\rangle$ with equal probability. With respect to a particular pair of orthogonal vectors $|a\rangle$ and $|b\rangle$, they result as

$$|\phi_\pm\rangle = \alpha|a\rangle \pm \beta|b\rangle,$$

where $\alpha = \sqrt{0.9}$ and $\beta = \sqrt{0.1}$. Design a quantum code that compresses three-qubit blocks into two qubits with fidelity up to 0.99.
5. For a source emitting mixed states, a better compression rate than the von Neumann entropy of the ensemble can be reached by knowing something about the produced states, i.e. by using a non-blind protocol. For such general protocols, an upper bound to the compression rate is known to be the minimum of the von Neumann entropy of the ensemble of purified source's states. Suppose we have a qubit source that generates two kinds of mixed states $\rho_1 = \begin{pmatrix} \frac{1}{2} & \frac{1}{4} \\ \frac{1}{4} & \frac{1}{2} \end{pmatrix}$ and $\rho_2 = \begin{pmatrix} \frac{3}{4} & 0 \\ 0 & \frac{1}{4} \end{pmatrix}$ with probabilities $p_1 = \frac{2}{5}$ and $p_2 = \frac{3}{5}$, respectively. Find the upper bound to the compression rate and compare it to $S(p_1\rho_1 + p_2\rho_2)$.

Chapter 9

INFORMATION TRANSMISSION

We describe here the most relevant classical and quantum communication scenarios and provide for them intuitive upper bounds on the information transmission capability. In Chapter 11, we will then show the achievability of these bounds through tools that will be developed in Chapter 10.

9.1 Classical and quantum Fano inequality

Suppose that a classical source X is sent by Alice through a noisy channel and Bob receives the variable Y. His task is to infer X by making a guess $\hat{X} = f(Y)$. It turns out that he is limited by the conditional entropy $H(X|Y)$.

Theorem 9.1.1 (Classical Fano inequality). *Suppose $\tilde{X} = f(Y)$ is used as guess of X and let $p_e := p(X \neq \tilde{X})$. Then,*

$$H(X|Y) \leq H_2(p_e) + p_e \log\left(|\mathcal{X}| - 1\right).$$

Proof. Define an error random variable E as taking value 1 if $X = \tilde{X}$ and value 0 if $X \neq \tilde{X}$. Then $H(E) \equiv H_2(p_e)$. Now observe that

$$\begin{aligned} H(E,X|Y) &= H(E,X,Y) - H(Y) \\ &= H(E,X,Y) - H(X,Y) + H(X,Y) - H(Y) \\ &= H(E|X,Y) + H(X|Y). \end{aligned} \tag{9.1}$$

But E is completely determined once X and Y are known, hence $H(E|X,Y) = 0$. Analogously, we can observe that

$$\begin{aligned} H(E,X|Y) &= H(E,X,Y) - H(Y) \\ &= H(E,X,Y) - H(E,Y) + H(E,Y) - H(Y) \\ &= H(X|E,Y) + H(E|Y). \end{aligned} \tag{9.2}$$

Since conditioning reduces entropy

$$H(E|Y) \leq H(E) = H_2(p_e),$$

whence from (9.1) and (9.2)

$$H(X|Y) = H(E,X|Y) \leq H(X|E,Y) + H_2(p_e).$$

Finally, observe that

$$\begin{aligned} H(X|E,Y) &= p(E=0)H(X|E=0,Y) + p(E=1)H(X|E=1,Y) \\ &\leq (1-p_e) \times 0 + p_e \times \log(|\mathcal{X}|-1). \end{aligned}$$

□

Suppose now Alice wishes to send a quantum source Q characterized by a density operator ρ through a quantum channel $\mathcal{N}$. First, we ask how much noise does $\mathcal{N}$ cause when applied to ρ. One measure of that is the extent to which the purification $|\psi\rangle$ of ρ (in the composite system RQ) becomes mixed.

Definition 9.1.2. The entropy exchange is defined as

$$S_e(\rho,\mathcal{N}) := S(R',Q') = S((\mathrm{id}_R \otimes \mathcal{N})(|\psi\rangle_{RQ}\langle\psi|)).$$

Theorem 9.1.3 (Quantum Fano inequality). *Given a density operator ρ and a quantum channel $\mathcal{N}$, it holds*

$$S_e(\rho,\mathcal{N}) \leq H_2\left(F(\rho,\mathcal{N})\right) + (1 - F(\rho,\mathcal{N}))\log(d^2-1),$$

where $F(\rho,\mathcal{N})$ is the entanglement fidelity and $d = |\mathcal{H}_Q|$.

Proof. Let $\{|i\rangle\}_i$ be an orthonormal basis for the system RQ such that the first state is $|1\rangle = |\psi_\rho\rangle_{RQ}$ the purification of ρ. If we write $\tilde{\rho}_{R'Q'} := \sum_i P_i\, \rho_{R'Q'}\, P_i$ with $P_i := |i\rangle\langle i|$, then $\tilde{\rho}_{R'Q'} = \sum_i p_i P_i$ where

$p_i := \langle i|\rho_{R'Q'}|i\rangle$. It follows (simply considering $D(\rho_{R'Q'}\|\tilde{\rho}_{R'Q'}) \geq 0$, see also Exercise 1, Chapter 5) that $S(\tilde{\rho}_{R'Q'}) \geq S(\rho_{R'Q'})$, i.e.

$$S(R', Q') \leq H(p_1, \ldots, p_{d^2}).$$

Moreover,

$$H(p_1, \ldots, p_{d^2}) = H_2(p_1) + (1-p_1)H\left(\frac{p_2}{1-p_1}, \ldots, \frac{p_{d^2}}{1-p_1}\right),$$

and in turn

$$H\left(\frac{p_2}{1-p_1}, \ldots, \frac{p_{d^2}}{1-p_1}\right) \leq \log(d^2-1).$$

Finally, taking into account that $p_1 \equiv F(\rho, \mathcal{N})$, we arrive at

$$S(\rho, \mathcal{N}) \leq H_2\left(F(\rho, \mathcal{N})\right) + (1 - F(\rho, \mathcal{N}))\log(d^2-1).$$

□

9.2 Classical information transmission

Suppose Alice wishes to communicate to Bob the symbols emitted by a classical source with corresponding random variable X. Let Y be the random variable received by Bob upon the action of a channel N. Then we ask: how much information can Bob *access* about X? The upper bound to the information Bob can access about X by knowing Y is clearly given by the mutual information $I(X:Y)$.

Now suppose that Alice encodes the symbol x into a suitable quantum state ρ_x and sends it through a quantum channel $\mathcal{N}$. Bob upon receiving it does a measurement by a POVM with outcome y, and tries to guess the symbol x. If we denote by Y the random variable corresponding to the outcomes of Bob's POVM, then his accessible information about X will be limited by $I(X:Y)$. However, this time the variable Y depends on the choice of the POVM made by Bob. Hence, the accessible information should be bounded by maximizing $I(X:Y)$ over all possible POVM.

Actually, an upper bound on the accessible information can be found in a different way.

Theorem 9.2.1 (Holevo bound). *Suppose Alice has a classical source $X \sim p_X$ over an alphabet $\mathcal{X}$. She encodes symbols x into ρ_x*

and sends them to Bob through a quantum channel $\mathcal{N}$. In turn, he performs measurement by POVM $\{E_y\}$. Let Y be the random variable corresponding to the outcomes of this POVM, then

$$I(X:Y) \leq \chi(\{p_x, \mathcal{N}(\rho_x)\}),$$

with the Holevo information

$$\chi(\{p_x, \mathcal{N}(\rho_x)\}) := S(\mathcal{N}(\rho)) - \sum_x p(x) S(\mathcal{N}(\rho_x)),$$

where $\rho := \sum_x p(x)\rho_x$.

Proof. Let $\mathcal{H}_A = span\{|x\rangle | x \in \mathcal{X}\}$ be the space associated with the messages of Alice and $\mathcal{H}_Q$ the space associated with the system where Alice encodes her messages. Finally, let $\mathcal{H}_B$ be the space associated with Bob's measurement device.

The initial state of the three systems reads

$$\rho_{AQB} = \sum_x p_x |x\rangle\langle x| \otimes \rho_x \otimes |0\rangle\langle 0|. \tag{9.3}$$

At the output of the channel, it becomes

$$\rho_{A'Q'B'} = \sum_x p_x |x\rangle\langle x| \otimes \mathcal{N}(\rho_x) \otimes |0\rangle\langle 0|.$$

Finally, after Bob's measurement chracterized by POVM $\{E_y\}$, it transforms to

$$\rho_{A''Q''B''} = \sum_{x,y} p_x |x\rangle\langle x| \otimes \sqrt{E_y}\mathcal{N}(\rho_x)\sqrt{E_y} \otimes |y\rangle\langle y|.$$

First, note that

$$I(A';Q') = I(A';Q',B'),$$

since prior the measurement B is uncorrelated with the rest. Second, we have

$$I(A';Q',B') \geq I(A'';Q'',B''),$$

since applying a CPTP map (that results from POVM $\{E_y\}$) to QB cannot increase the quantum mutual information with A. Third, we note that

$$I(A''; Q'', B'') \geq I(A''; B''),$$

since discarding a system cannot increase the quantum mutual information. Putting these together, we obtain the inequality:

$$I(A''; B'') \leq I(A'; Q'), \tag{9.4}$$

which correpsonds exactly to the Holevo bound. To see this explicitly, let us evaluate separately the l.h.s. and the r.h.s. of (9.4).

At the l.h.s. of (9.4), we notice that

$$\rho_{A''B''} = \mathrm{Tr}_{Q''}(\rho_{A''Q''B''}) = \sum_{x,y} p_x \, \mathrm{Tr}\,(\mathcal{N}(\rho_x)E_y)\,|x\rangle\langle x| \otimes |y\rangle\langle y|,$$

which is analogous to a classical joint probability distribution $p(x, y) = p(y|x)p(x)$ (since the states are orthonormal), where $p(y|x) = \mathrm{tr}\,(\mathcal{N}(\rho_x)E_y)$. This implies that $I(A''; B'') = I(X : Y)$.

At the r.h.s. of (9.4), we notice that

$$\rho_{A'Q'} = \sum_x p_x |x\rangle\langle x| \otimes \mathcal{N}(\rho_x),$$

$$\rho_{A'} = \sum_x p_x |x\rangle\langle x|,$$

$$\rho_{Q'} = \sum_x p_x \mathcal{N}(\rho_x),$$

and then (noticing that $|x\rangle\langle x| \otimes \rho_x$ have support on orthogonal subspace)

$$\begin{aligned} I(A'; Q') &= S(A') + S(Q') - S(A', Q') \\ &= H(p_x) + S(\mathcal{N}(\rho_Q)) - \left[H(p_x) + \sum_x p_x S\left(\mathcal{N}(\rho_x)\right) \right] \\ &= S\left(\mathcal{N}(\rho_Q)\right) - \sum_x p_x S\left(\mathcal{N}(\rho_x)\right). \end{aligned}$$

□

We now want to prove that furthermore

$$\chi(\{p_x, \mathcal{N}(\rho_x)\}) \leq H(X). \tag{9.5}$$

To this end let us first prove the following result.

Proposition 9.2.2. *Given* $\sigma = \sum_x p_x \sigma_x$ *and* x *realizations of the random variable* $X \sim p_x$ *on* $\mathcal{X}$*, it holds*

$$S(\sigma) \leq H(X) + \sum_x p_x S(\sigma_x). \tag{9.6}$$

Proof. Consider first the case of pure states $\sigma_x = |\psi_x\rangle\langle\psi_x|$. Introduce a reference system R with orthonormal basis $\{|x\rangle\}$ and define

$$|\Psi\rangle_{RQ} := \sum_x \sqrt{p_x}|x\rangle|\psi_x\rangle. \tag{9.7}$$

Since $|\Psi\rangle_{RQ}$ is pure, we have $S(\rho_R) = S(\sigma)$ with $\rho_R = \mathrm{Tr}_Q|\Psi\rangle_{RQ}\langle\Psi|$. Suppose now to perform a projective measurement on the system R in the basis $\{|x\rangle\}$ obtaining the state

$$\widetilde{\rho}_R = \sum_x p_x|x\rangle\langle x|. \tag{9.8}$$

for the system R. Since a projective measurement does not decrease the entropy, $S(\sigma) = S(\rho_R) \leq S(\widetilde{\rho}_R) = H(X)$. Given that $S(\sigma_x) = 0$ for pure states, we proved

$$S(\sigma) \leq H(X) + \sum_x p_x S(\sigma_x). \tag{9.9}$$

The case of mixed states σ_x immediately follows by considering the spectral decomposition of σ_x and applying the pure state result. □

Note that (9.6) relates the von Neumann entropy to the Shannon entropy, and it reduces to property (4) of Section 5.5 when the σ_x have support on mutually orthogonal subspaces.

Now if in Eq. (9.6) we consider $\sigma = \mathcal{N}(\rho) = \mathcal{N}(\sum_x p_x \rho_x)$ we immediately get, together with the Holevo bound, the following chain of inequalities:

$$I(X:Y) \leq \chi(\{p_x, \mathcal{N}(\rho_x)\}) \leq H(X). \tag{9.10}$$

It is then clear that one qubit can encode at most one bit of classical information. In general, given a quantum system with Hilbert space of dimension d, it can be used to encode at most $\log d$ bits of classical information reliably.

9.3 Quantum information transmission

Transmitting quantum information through a channel $\mathcal{N}$ means to transmit a quantum source characterized by a density operator ρ, or in other words the entanglement between the main system and a reference system. Then an upper bound on the entropy rate of the transmitted quantum source is provided by the following theorem.

Theorem 9.3.1. *Given a quantum source and a channel $\mathcal{N}$, it is*

$$S(\rho) \leq I_{\mathrm{coh}}(\rho, \mathcal{N}) + 2H_2\left(F(\rho, \mathcal{N})\right) + 2\left(1 - F(\rho, \mathcal{N})\right)\log(d^2 - 1).$$

Proof. By the definition of coherent information 6.4.1., we have

$$S(\rho) - I_{\mathrm{coh}}(\rho, \mathcal{N}) = S(\rho) - S(\mathcal{N}(\rho)) + S(\rho, \mathcal{N}),$$

where the last term is the entropy exchange 9.1.1. Then

$$S(\rho) - I_{\mathrm{coh}}(\rho, \mathcal{N}) \leq 2S(\rho, \mathcal{N}), \tag{9.11}$$

because

$$\begin{aligned} S(\rho) - S(\mathcal{N}(\rho)) &= S(R) - S(Q') = S(R') - S(Q') \\ &\leq S(R', Q') = S(E') = S(\rho, \mathcal{N}). \end{aligned}$$

Finally using the quantum Fano inequality, Theorem 9.1.3, in Eq. (9.11), we get the desired result. □

It is worth noticing that if the entanglement fidelity gets close to one, the upper bound is actually provided by the coherent information $I_{\mathrm{coh}}(\rho, \mathcal{N})$.

9.4 Entanglement-assisted information transmission

Let us first examine two paradigmatic protocols where entanglement pre-shared by Alice and Bob is exploited for communicating.

9.4.1 *Teleportation*

We describe the protocol for the case of a qubit (the generalization to the case of Hilbert spaces with higher dimensions is straight forward).

- Consider a two-qubit maximally entangled state vector, e.g.

$$|\Phi^+\rangle = \frac{|00\rangle + |11\rangle}{\sqrt{2}}, \tag{9.12}$$

where the first qubit belongs to Alice and the second to Bob. Alice aims to communicate an unknown qubit state vector $|\psi\rangle = a|0\rangle + b|1\rangle$ to Bob.

Then, the initial state vector of the three qubits system (qubit to be teleported, pairs of entangled qubits shared between Alice and Bob) reads

$$|\psi\rangle|\Phi^+\rangle = (a|0\rangle + b|1\rangle)\,\frac{|00\rangle + |11\rangle}{\sqrt{2}}, \tag{9.13}$$

which by using Bell's state vectors (see (3.8)) on Alice's side can be rewritten as

$$\begin{aligned}|\psi\rangle|\Phi^+\rangle =& \frac{1}{2}[|\Phi^+\rangle\,(a|0\rangle + b|1\rangle) + |\Psi^+\rangle\,(a|1\rangle + b|0\rangle)\\ &+ |\Phi^-\rangle\,(a|0\rangle - b|1\rangle) + |\Psi^-\rangle\,(a|1\rangle - b|0\rangle)].\end{aligned} \tag{9.14}$$

- Alice performs a projective measurement on the Bell basis (with POVM elements $E_{00} = |\Phi^+\rangle\langle\Phi^+|$, $E_{01} = |\Psi^+\rangle\langle\Psi^+|$, $E_{10} = |\Phi^-\rangle\langle\Phi^-|, E_{11} = |\Psi^-\rangle\langle\Psi^-|$). Depending on the result of this measurement, the state vector of Bob's qubit will collapse according to

$$\text{Alice's result } 00 \rightarrow \text{Bob's state } a|0\rangle + b|1\rangle, \tag{9.15}$$

$$\text{Alice's result } 01 \rightarrow \text{Bob's state } a|1\rangle + b|0\rangle, \tag{9.16}$$

$$\text{Alice's result } 10 \rightarrow \text{Bob's state } a|0\rangle - b|1\rangle, \tag{9.17}$$

$$\text{Alice's result } 11 \rightarrow \text{Bob's state } a|1\rangle - b|0\rangle. \tag{9.18}$$

Each of these results happens with probability 1/4.

- Alice sends the measurement result to Bob. This requires a classical communication channel able to send two bits of classical information.
- In the final step of the protocol, Bob, according to the information received, applies a unitary transformation on his qubit:

$$\text{if } 00 \text{ is received, Bob applies } U_{00} = I, \tag{9.19}$$

$$\text{if } 01 \text{ is received, Bob applies } U_{01} = \sigma^X, \tag{9.20}$$

$$\text{if } 10 \text{ is received, Bob applies } U_{10} = \sigma^Z, \tag{9.21}$$

$$\text{if } 11 \text{ is received, Bob applies } U_{11} = \sigma^Z \sigma^X. \tag{9.22}$$

At the end this stage, the qubit of Bob results in the same state vector $|\psi\rangle$ of Alice's initial qubit.

It is worth noting that in this protocol it is the "quantum information" which moves from Alice's qubit to Bob's qubit. No matter or energy is teleported, but only quantum information.

Moreover, due to Alice's measurement, the initial state is lost (in accordance with the no-cloning theorem). For the same reason, there is no more shared entanglement between Alice and Bob at the end of the protocol.

In conclusion, one qubit of quantum information has been sent from Alice to Bob. In order to do that, two bits of classical information have been sent from Alice to Bob and a maximally entangled pair of qubits has been used.

In this way, quantum communication over an ideal quantum channel has been simulated. Whenever the resources used are not perfect (e.g. shared entangled is not maximal or classical communication takes place with errors), we will end up with quantum communication over a noisy quantum channel, for which an upper bound has been found in the previous section.

Measure-and-prepare teleportation protocol

One could ask whether pre-shared entanglement is a needed resource to simulate the quantum channel. Alternatively, one can try to send quantum information by using only local measurements by Alice and local transformations by Bob, and classical information from Alice to Bob. These kinds of schemes are called *measure and prepare* protocols.

In the case of a qubit, given an initial (unknown) state vector $|\psi\rangle = a|0\rangle + b|1\rangle$, the best that Alice can do is to measure the qubit on a basis, then to send the resulting classical information to Bob, who prepares the state of his qubit accordingly.

Choosing randomly the basis over which Alice measures and keeps fixed the state vector $|\psi\rangle$ is equivalent to always measuring the same observable, say σ^Z, and randomly varying the coefficients $a = \cos(\theta/2)$ and $b = e^{i\varphi}\sin(\theta/2)$ according to the probability measure $(1/4\pi)d\varphi \sin\theta d\theta$.

Then, on each round Alice will get 0 or 1 with probability $|a|^2$, $|b|^2$. If she gets 0 (resp., 1), she sends this information to Bob, who can at its best prepare the state $|0\rangle\langle 0|$ (resp., $|1\rangle\langle 1|$): the resulting fidelity is $F(|\psi\rangle, |0\rangle) = |a|^2$ (resp., $F(|\psi\rangle, |1\rangle) = |b|^2$). The average fidelity is then

$$\begin{aligned}\overline{F} &= \overline{|a|^4 + |b|^4} \\ &= \frac{1}{4\pi}\int_0^{2\pi} d\phi \int_0^{\pi} \left[\cos^4(\theta/2) + \sin^4(\theta/2)\right]\sin\theta d\theta = \frac{2}{3}. \qquad (9.23)\end{aligned}$$

That is, when using a "measure and prepare" protocol, quantum information can be sent only with limited fidelity < 1; for the case of a qubit, this cannot be larger than 2/3. Furthermore, it can be proven that all the measure and prepare protocols give zero coherent information.

9.4.2 *Superdense coding*

We are now going to analyze the converse goal with respect to the one achieved by quantum teleportation protocol, that is rather than sending one qubit by two classical bits with the help of a maximally entangled pair of qubits, we see how to send two bits by one qubit with the help of a maximally entangled pair of qubits (a protocol termed superdense coding).

- Consider two qubits in a maximally entangled state vector, e.g.

$$|\Phi^+\rangle = \frac{|00\rangle + |11\rangle}{\sqrt{2}}, \tag{9.24}$$

  where the first qubit belongs to Alice, and the second to Bob.
- Alice applies one of the four unitaries $U_{00} = I$, $U_{01} = \sigma^X$, $U_{10} = \sigma^Z$, $U_{11} = \sigma^X\sigma^Z$ to her qubit, thus obtaining

$$I \otimes I|\Phi^+\rangle = |\Phi^+\rangle, \tag{9.25}$$

$$\sigma^X \otimes I|\Phi^+\rangle = |\Psi^+\rangle, \tag{9.26}$$

$$\sigma^Z \otimes I|\Phi^+\rangle = |\Phi^-\rangle, \tag{9.27}$$

$$\sigma^X\sigma^Z \otimes I|\Phi^+\rangle = |\Psi^-\rangle. \tag{9.28}$$

  In this way, Alice can actually encode two bits of classical information.
- Alice sends her qubit to Bob through an ideal (noiseless) quantum channel, hence one qubit of quantum information is communicated from Alice to Bob.
- At this stage, Bob has two qubits and makes a projective measurement on the Bell basis determining (without ambiguity) which of the four Bell state vectors has been prepared by Alice.

In conclusion, Alice and Bob have consumed a pair of maximally entangled qubits, used one qubit of quantum communication, local measurement and transformations. What they have obtained at the end is two bits of classical communication (this is not a violation of the Holevo bound, since the superdense protocol uses an additional resource: preshared entanglement).

If the quantum channel used in superdense coding is noisy, and allowing an arbitrary shared state ω_{UV} and arbitrary quantum channels $\mathcal{E}_x : \mathfrak{D}(\mathcal{H}_U) \to \mathfrak{D}(\mathcal{H}_A)$ to encode information x, occurring with probability p_x, we obtain the following upper bound on the information accessible to Bob. To state it, let us introduce the encoding states $(\rho_x)_{AV} := (\mathcal{E}_x \otimes \mathrm{id})\omega$, and $\rho_A := \sum_x p_x \mathrm{Tr}_V \rho_x$.

Theorem 9.4.1. *Under the above notations, and with an arbitrary POVM of Bob's with outcomes y,*

$$I(X:Y) \leq S(\rho) + S(\mathcal{N}) - S\big((\mathcal{N} \otimes \mathrm{id})|\psi_\rho\rangle\langle\psi_\rho|\big), \tag{9.29}$$

where $|\psi_\rho\rangle$ is a purification (in AA') of ρ (belonging to A).

Proof. Bob receives the states $(\sigma_x)_{BV} = (\mathcal{N} \otimes \mathrm{id})\rho_x$, so by the Holevo bound, Theorem 9.2.1, we have

$$I(X:Y) \leq \chi(\{p_x, \sigma_x\}) = I(X:BV)_\sigma,$$

where the mutual information is with respect to the state

$$\sigma_{BVX} = \sum_x p_x (\sigma_x)_{BV} \otimes |x\rangle\langle x|_X = (\mathcal{N} \otimes \mathrm{id}_{VX})\Omega_{AVX},$$

and we have introduced

$$\Omega_{AVX} := \sum_x p_x (\rho_x)_{AV} \otimes |x\rangle\langle x|_X.$$

The latter state can be purified to $|\Omega\rangle_{AVXE}$, and because $\Omega_A = \rho_A$, by the purification Theorem 4.3.2, this means that there exists an isometry $\mathcal{V} : \mathcal{H}_{A'} \to \mathcal{H}_V \otimes \mathcal{H}_X \otimes \mathcal{H}_E$ with $(I \otimes \mathcal{V})|\psi\rangle = |\Omega\rangle$.

Now, we can use the formula for the quantum mutual information, the subadditivity of the entropy and the chain rule to obtain

$$\begin{aligned} I(X:BV) &= S(BV) - S(BV|X) \\ &\leq S(B) + S(V) - S(V|X) - S(B|VX) \\ &= S(B) - S(B|VX) = I(B:VX), \end{aligned}$$

where we have exploited $S(V) - S(V|X) = I(X:V) = 0$, which comes from the fact that for all x, $(\rho_x)_V = (\sigma_x)_V = \omega_V$ since the channels $\mathcal{E}_x$ and $\mathcal{N}$ act on U.

But then, by strong subadditivity of the entropy,

$$I(B:VX)_\sigma \leq I(B:VXE)_{(\mathcal{N}\otimes \mathrm{id}_{VXE})\Omega_{AVXE}} = I(B:A')_{(\mathcal{N}\otimes \mathrm{id}_{A'})\psi_{AA'}},$$

the latter by the unitary invariance of entropies. We conclude by noting that here, $I(B:A')$ equals the r.h.s. of Eq. (9.29). □

Exercises

1. Study the conditions under which the classical (resp. quantum) Fano inequality becomes an equality.
2. Prove that the entropy exchange $S_e(\rho, \mathcal{N})$ is concave with respect to both arguments.
3. Suppose that Alice sends the following four state vectors to Bob each with probability 1/4:

$$\begin{aligned}
|x_1\rangle &= |0\rangle, \\
|x_2\rangle &= \frac{1}{\sqrt{3}}(|0\rangle + \sqrt{2}|1\rangle), \\
|x_3\rangle &= \frac{1}{\sqrt{3}}(|0\rangle + \sqrt{2}e^{2\pi i/3}|1\rangle), \\
|x_4\rangle &= \frac{1}{\sqrt{3}}(|0\rangle + \sqrt{2}e^{4\pi i/3}|1\rangle).
\end{aligned}$$

   Evaluate the Holevo bound and design a POVM that allows to achieve a mutual information greater than 0.4 bit.
4. Find the fidelity that can be attained to teleport a qubit from Alice to Bob when they share the following qubit pair state:

$$\rho_{AB}(\lambda) = (1-\lambda)|\Phi^{(+)}\rangle\langle\Phi^{(+)}| + \frac{\lambda}{4} I_{AB},$$
$$|\Phi^{(+)}\rangle = \frac{1}{\sqrt{2}}\left(|00\rangle + |11\rangle\right),$$

   with $\lambda \in [0,1]$. For what values of λ does it overcome the upper bound $F = 2/3$ achievable classically?

5. Show that the following qubit pair state:

$$\rho_{AB}(\lambda) = (1-\lambda)|\Phi^+\rangle\langle\Phi^+| + \frac{\lambda}{2}\left(|00\rangle\langle 00| + |01\rangle\langle 01|\right),$$

with $0 < \lambda \leq 1$, is entangled but not useful for dense coding.

Chapter 10

ERROR CORRECTING CODES

To transmit information through a noisy channel in a reliable way, one has to deal with the effect of errors. It is customary to counteract such unwanted effects by exploiting redundancy. This is the subject of the error correction theory that we present here, by focusing for the sake of simplicity only on bits and qubits.

10.1 Basics of classical error correction theory

Example. Suppose that Alice sends one bit of information to Bob via a symmetric binary channel with flipping-probability equal to p. A simple error correcting code they can use is known as *repetition code.* It works as follows. When Alice wants to send a bit value "0" or "1" to Bob, instead of sending it directly, she makes three copies of it. This defines a map from the value of one bit to the value of three bits:

$$0 \to 000,$$
$$1 \to 111.$$

Let us now see the effects of the errors assuming that they occur on different bits and are independent and identically distributed.

- With probability $(1-p)^3$, no bit-flip occurs, hence mapping

$$000 \to 000$$
$$111 \to 111$$

- With probability $p(1-p)^2$, only one bit-flip occurs, hence mapping

$$000 \to 100 \text{ or } 010 \text{ or } 001$$
$$111 \to 011 \text{ or } 101 \text{ or } 110$$

- With probability $p^2(1-p)$, two bit-flips occur, hence mapping

$$000 \to 011 \text{ or } 101 \text{ or } 110$$
$$111 \to 100 \text{ or } 010 \text{ or } 001$$

- With probability p^3, three bit-flips occur, hence mapping

$$000 \to 111$$
$$111 \to 000$$

Bob decodes by *majority voting*, i.e. a string of three bits is decoded as 0 (resp., 1) if the majority of symbols there are zeros (resp., ones).

Clearly, this way of decoding is not perfect. Indeed, an error occurs whenever 2 or 3 single bit errors occurred. Hence, the probability of error using the *repetition code* and *majority vote* decoding is given by

$$p_e = 3p^2(1-p) + p^3.$$

In order to have an effective correcting scheme, we must require $p_e < p$ (p as the probability of error without any error correcting code). It is easy to see that this holds true for $p < 1/2$ (clearly, if $p > 1/2$ one can get an effective code by modifying the previous encoding, so to have $0 \to 111$ and $1 \to 000$ and use the same-majority voting-decoding procedure).

Starting from the above example, we can outline the general theory of error correction.

Definition 10.1.1. Given an encoding map and a decoding map

$$E : \mathbb{F}_2^k \to \mathbb{F}_2^n, \quad \Delta : \mathbb{F}_2^n \to \mathbb{F}_2^k, \tag{10.1}$$

with $n > k$, the code Γ is the image of E, that is a subset of the

Hamming space $\mathbb{F}_2^n$. The elements of the code are called codewords and the code rate is the ratio k/n.

The effect of a noisy channel is to transform a codeword u into another vector u'. The difference $e = u' - u$ is the *error* induced by the channel. The aim of error correction is to deduce the error e from the output u'. If Bob is able to do that, he can recover the original codeword: $u = u' + e$.

Theorem 10.1.2. *Given a code $\Gamma \subseteq \mathbb{F}_2^n$, the set of errors that can be corrected are those for which $u + e^{(i)} \neq v + e^{(j)}$, $\forall\ u \neq v,\ u, v \in \Gamma$. In other words,* $\text{wt}(u + v + e^{(i)} + e^{(j)}) \neq 0$.

The proof is trivial.

For the three-fold repetition code, the correctable errors are $\{000, 001, 010, 100\}$. These are the most probable errors for $p < 1/2$. That is why this code is effective for $p < 1/2$. This code can correct errors with at most Hamming weight equal to 1.

Definition 10.1.3. Given a code $\Gamma \subseteq \mathbb{F}_2^n$, the minimum distance of the code is the minimum Hamming distance between the codewords.

Theorem 10.1.4. *A code of minimum distance d can correct all errors of weight up to t if and only if $d > 2t$.*

Proof. If the code can correct all errors of weight up to t, that means that $\text{wt}(u + v + e^{(i)} + e^{(j)}) > 0$ for all u and v in the code book and for all errors $e^{(i)}$, $e^{(j)}$ with weight less or equal to t. That implies $\text{wt}(u + v) > 2t$ for all codewords u and v. Hence, $d > 2t$.

Vice versa, if $d > 2t$, it means that for all codewords $\text{wt}(u + v) > 2t$. Since for all errors $e^{(i)}$, $e^{(j)}$ with weight up to t $\text{wt}(e^{(i)} + e^{(j)}) \leq 2t$, this implies $\text{wt}(u + v + e^{(i)} + e^{(j)}) > 0$ for all codewords and all errors with weight up to t. □

Note that this theorem does not forbid the existence of some correctable errors of weight bigger than $d/2$.

A good code should have the biggest possible d and also the highest possible rate k/n.

If Alice and Bob agree to use a certain code, in order to correct the errors, they should write a table containing all the erroneous vectors u' for each codeword u and all possible errors. Such a table can be used by Bob to recognize the correctable errors and correct them. This can always be done in principle, but is highly inefficient with increasing n. A more efficient procedure makes use of the properties of *linear codes.*

10.2 Classical linear codes

Definition 10.2.1. A linear code Γ is a code which is a linear subspace of the Hamming space $\mathbb{F}_2^n$, i.e. it contains the null vector and is closed under vector addition.

Given a basis for this subspace Γ (i.e. a set of linearly independent generators) $\{u^{(1)}, \ldots, u^{(k)}\}$, the $k \times n$ matrix

$$\mathsf{G} = \begin{pmatrix} u^{(1)} \\ u^{(2)} \\ \ldots \\ u^{(k)} \end{pmatrix} \tag{10.2}$$

is called the generator matrix.

A linear code of 2^k elements, made of strings of n bits, with a minimal Hamming distance d is denoted as $[n, k, d]$-code.[1]

Example. The three-fold repetition code is $\{000, 111\} \subset \mathbb{F}_2^3$ and $\mathsf{G} = (1\,1\,1)$, thus defining a $[3, 1, 3]$-code.

Given a linear subspace $\Gamma \subset \mathbb{F}_2^n$ with generator matrix G, we can consider the dual subspace $\Gamma^\perp$, defined as the subset of orthogonal vectors. It is a linear subspace too. It happens that if $|\Gamma| = k$, then $|\Gamma^\perp| = n - k$.

Definition 10.2.2. The generator matrix of the dual space $\Gamma^\perp$ is called *parity-check matrix* and denoted by H.

---

[1]For a generic linear code without specified minimum, distance, we will use the notation $[n, k]$.

If G is a $k \times n$ matrix, H is a $(n-k) \times n$ matrix. The orthogonality condition reads

$$\mathsf{H}\mathsf{G}^\top = 0. \tag{10.3}$$

Note that, since the scalar product is degenerate, it could be that $\mathsf{G} = \mathsf{H}$, in which case the code is said to be self-dual.

Two codes are *equivalent* if the generator matrix of one of them can be obtained from the other by permutation of the columns. It follows that the generator matrix (up to equivalence) can always be written in the form

$$\mathsf{G} = \left(\mathsf{I}_{k\times k}\ \mathsf{A} \right), \tag{10.4}$$

where $\mathsf{I}_{k\times k}$ is the $k \times k$ identity matrix, and A is a $k \times (n-k)$ matrix. Then, the parity-check matrix results

$$\mathsf{H} = \left(\mathsf{A}^\top\ \mathsf{I}_{(n-k)\times(n-k)} \right). \tag{10.5}$$

Example. For the three-fold repetition code,

$$\mathsf{G} = (1\,1\,1) \Rightarrow \mathsf{H} = \begin{pmatrix} 1\,1\,0 \\ 1\,0\,1 \end{pmatrix}.$$

The parity-check matrix has the following property:

$$\mathsf{H}u^\top = 0 \tag{10.6}$$

if and only if u belongs to the (linear) code.

Given an input codeword u, if an (unknown) error e happens, the output is $u' = u + e$. By linearity, we have

$$\mathsf{H}(u')^\top = \mathsf{H}(u+e)^\top = \mathsf{H}e^\top. \tag{10.7}$$

Definition 10.2.3. The vector $\mathsf{H}e^\top$ is called the error syndrome (it is only a function of the — unknown — error and is independent of the input u).

Note that there exists at most 2^{n-k} different error syndromes (while there are 2^n erroneous vectors). It is also worth noting that the elements of the syndrome vector are the parity of the error vector with respect to the row of the parity-check matrix.

Example. For the three-fold repetition code,

$$\begin{pmatrix} 1\,1\,0 \\ 1\,0\,1 \end{pmatrix} \begin{pmatrix} 1 \\ 0 \\ 0 \end{pmatrix} = \begin{pmatrix} 1 \\ 1 \end{pmatrix}, \quad \begin{pmatrix} 1\,1\,0 \\ 1\,0\,1 \end{pmatrix} \begin{pmatrix} 0 \\ 1 \\ 0 \end{pmatrix} = \begin{pmatrix} 1 \\ 0 \end{pmatrix},$$

$$\begin{pmatrix} 1\,1\,0 \\ 1\,0\,1 \end{pmatrix} \begin{pmatrix} 0 \\ 0 \\ 1 \end{pmatrix} = \begin{pmatrix} 0 \\ 1 \end{pmatrix}, \quad \begin{pmatrix} 1\,1\,0 \\ 1\,0\,1 \end{pmatrix} \begin{pmatrix} 0 \\ 0 \\ 0 \end{pmatrix} = \begin{pmatrix} 0 \\ 0 \end{pmatrix}.$$

The parity-check matrix is composed of $n - k$ parity-check vectors. Each of them divides the Hamming space in two parts: vectors with corresponding parity equal 0 or 1 (see Section 1). Hence, there are 2^k vectors in the Hamming space having the same syndrome

$$\frac{2^n}{\underbrace{2 \times 2 \times \ldots \times 2}_{n-k \; times}} = 2^k.$$

On the other hand, we know that $\mathsf{H}e^\top = \mathsf{H}(u + e)\top$, $\forall u \in \Gamma$. Since there are 2^k codewords, the 2^k vectors with syndrome $\mathsf{H}e^\top$ are the vectors $u + e$, for all $u \in \Gamma$. In conclusion, there exists only one correctable error for each error syndrome. So, what is needed is a look-up table connecting each syndrome with its (unique) correctable error.

10.2.1 *Bounds on classical codes*

Each of the 2^k codewords has a Hamming sphere of radius t.[2] All the words inside the Hamming sphere come from errors acting on the same codeword. For a code on n bits, there are n one-bit errors, $\binom{n}{2}$ two-bit errors, and in general $\binom{n}{j}$ j-bit errors. The Hamming spheres

[2]Given a vector v on the Hamming space $\mathbb{F}_2^n$, a (Hamming) sphere centered on it of radius t is the set of vectors lying within a Hamming distance smaller or equal to t with respect to v.

cannot overlap, but they must fit inside the vector space $\mathbb{F}_2^n$, which only contains 2^n elements. Thus,

$$2^k \sum_{i=0}^{t} \binom{n}{i} \leq 2^n . \qquad (10.8)$$

This is the so-called *Hamming bound.*

Codes that saturate the Hamming bound are called *perfect codes.*

Example. An example of perfect codes is provided by the Hamming codes $[2^r-1, 2^r-r-1, 3]$ with integer $r \geq 2$. The parity-check matrix of such codes is build up by considering as columns all possible strings of r bit, but the null one. The case of $r = 3$ gives a $[7, 4, 3]$ code with

$$\mathsf{H} = \begin{pmatrix} 0 & 0 & 0 & 1 & 1 & 1 & 1 \\ 0 & 1 & 1 & 0 & 0 & 1 & 1 \\ 1 & 0 & 1 & 0 & 1 & 0 & 1 \end{pmatrix} . \qquad (10.9)$$

Taking the log on both sides of (10.8) gives

$$k \leq n - \log \left[\sum_{i=0}^{t} \binom{n}{i} \right] ,$$

and applying the Stirling approximation $(\log(x!) \sim x \log(x) - x)$ for large n and t/n fixed yields

$$\frac{k}{n} \lesssim 1 - H_2(t/n). \qquad (10.10)$$

Now suppose Γ to be an $[n, k, d]$ code. If we delete the last $d-1$ coordinates from each codeword, then the 2^k vectors of length $n-d+1$ so obtained must be distinct (given that the minimum distance in the code is d), hence

$$k \leq n - d + 1. \qquad (10.11)$$

This is known as *Singleton bound.*

Both Hamming and Singleton bounds are upper bound on the rate achievable by correcting codes. We can set a lower bound on the existence of $[n, k, 2t+1]$ linear codes as well. This will show that good codes actually exist.

Suppose we have a code (if necessary with $k = 0$) with

$$\sum_{j=0}^{2t} \binom{n}{j} 2^k < 2^n.$$

Then the spheres of distance $2t$ around each codeword do not fill the space, so there is some vector v that is at least at distance $2t+1$ from each of the other codewords. In addition, $v+s$ (for any codeword s) is at least at distance $2t+1$ from any other codeword s', since the distance is just $(v+s)+s' = v+(s+s')$, which is the distance between v and the codeword $s+s'$. This means that we can add v and all the vectors $v+s$ to the code without dropping the distance below $2t+1$. This gives us an $[n, k+1, 2t+1]$ code. We can continue this process until

$$\sum_{j=0}^{2t} \binom{n}{j} 2^k \geq 2^n.$$

In other words, we will have the *Gilbert–Varshamov bound*

$$k \geq \max\left\{ k' \middle| 2^{k'} \sum_{i=0}^{2t} \binom{n}{i} < 2^n \right\}. \tag{10.12}$$

For large n and t/n fixed, this yields

$$\frac{k}{n} \gtrsim 1 - H_2(2t/n). \tag{10.13}$$

10.3 From classical to quantum codes

Moving to the quantum framework, we start analyzing the quantum analogue of the three-fold repetition code given by the quantum codewords

$$|0_L\rangle = |000\rangle, \quad |1_L\rangle = |111\rangle. \tag{10.14}$$

The two-dimensional space generated by $|0_L\rangle$, $|1_L\rangle$ defines a so-called *logical qubit*, which encodes quantum information.

We expect that this quantum code can be used to correct *bit-flip* errors introduced by the CPTP map

$$\mathcal{N}(\rho) = (1-p)\rho + p\,\sigma^X \rho \sigma^X \,. \qquad (10.15)$$

Since quantum states cannot be copied, the first problem is to find a unitary transformation for the encoding (using ancillary systems).

We start from a state vector $|\psi\rangle = \alpha|0\rangle + \beta|1\rangle$ and add two ancilla systems in the state vector $|0\rangle|0\rangle$ yielding $|\psi\rangle|0\rangle|0\rangle$. Then, we apply

$$U^{(13)}_{CNOT} U^{(12)}_{CNOT} |\psi\rangle|0\rangle|0\rangle = \alpha|000\rangle + \beta|111\rangle \,, \qquad (10.16)$$

obtaining the desired state vector for any α and $\beta \in \mathbb{C}$.

Now we have to devise a suitable measurement strategy to extract the error syndrome. Suppose we measure $\sigma^Z_1 \otimes \sigma^Z_2$ and $\sigma^Z_2 \otimes \sigma^Z_3$, then we have

outcome	error
$+1\,+1$	no error
$-1\,+1$	error on first qubit
$+1\,-1$	error on third qubit
$-1\,-1$	error on second qubit

It is clear that this POVM does not acquire information on the state, but only on the occurred error. Then, a single bit-flip error can be undone by applying the unitary σ^X on the qubit on which the bit-flip error was located.

Both syndrome extraction and error correction are completely positive maps (but not necessarily trace-preserving). Actually, it is not necessary to distinguish them, one could simply define a single CPTP map (the *recovery* $\mathcal{R}$) which includes both.

Similarly to the classical case, we get that the probability of error is reduced to $p_e = 3p^2 - 2p^3$, which is smaller than p for $p < 1/2$.

However, in the quantum framework, a more appropriate comparison is done in terms of the fidelity

$$F = \sqrt{\langle\psi|\mathcal{N}(|\psi\rangle\langle\psi|)|\psi\rangle} = \sqrt{(1-p) + p|\langle\psi|\sigma^X|\psi\rangle|^2}, \qquad (10.17)$$

which reaches its minimum $\sqrt{1-p}$ for $|\psi\rangle = |0\rangle$ or $|\psi\rangle = |1\rangle$.

After error correction, the original state is recovered with probability $1 - p_e$, hence the post-error correction state will be

$$\rho = (1 - p_e)|\psi\rangle\langle\psi| + \cdots, \tag{10.18}$$

where "..." stands for a sum of positive operators corresponding to the un-corrected terms. Then, the fidelity reads

$$F = \sqrt{\langle\psi|\rho|\psi\rangle} = \sqrt{(1 - p_e) + \text{positive terms}} \geq \sqrt{1 - p_e}, \tag{10.19}$$

and its minimum results greater than the previous one $\sqrt{1-p}$ for $p < 1/2$.

The phase-flip errors can be corrected analogously by noticing that the phase-flip map is unitarily equivalent to the bit-flip one

$$\mathcal{N}_{\text{phase-flip}}(\rho) = H\mathcal{N}_{\text{bit-flip}}(H\rho H^\dagger)H, \tag{10.20}$$

where H is the Hadamard operator.

The repetition code of the phase-flip noise is hence given by the quantum codewords

$$|{+}{+}{+}\rangle, \quad |{-}{-}{-}\rangle, \tag{10.21}$$

where $|\pm\rangle$ are the eigenvectors of the Pauli operator σ^X defined in 3.6.

Given a state vector $|\psi\rangle = \alpha|0\rangle + \beta|1\rangle$, the encoding map (10.16) transforms to

$$H^{\otimes 3} U_{CNOT}^{(13)} U_{CNOT}^{(12)} |\psi\rangle|0\rangle|0\rangle = \alpha|{+}{+}{+}\rangle + \beta|{-}{-}{-}\rangle\,. \tag{10.22}$$

Similarly, the error detection can be made by measuring the operators $H^{\otimes 3}\sigma_1^Z \otimes \sigma_2^Z H^{\otimes 3} = \sigma_1^X \otimes \sigma_2^X$, and $H^{\otimes 3}\sigma_2^Z \otimes \sigma_3^Z H^{\otimes 3} = \sigma_2^X \otimes \sigma_3^X$. Then, a single phase-flip error can be undone by applying the unitary σ^Z on the qubit on which the phase-flip error was located.

Quite generally, both bit-flip and phase-flip errors may occur in the same qubit, hence it would be desirable to also correct bit-phase-flip errors.

This can be done by *concatenating* the bit-flip and the phase-flip repetition codes (the code thus obtained is named Shor code). It works as follows:

First encode $|0\rangle$ and $|1\rangle$ as

$$|0\rangle \to |{+}{+}{+}\rangle, \quad |1\rangle \to |{-}{-}{-}\rangle,$$

then encode $|0\rangle$ and $|1\rangle$ present into $|+\rangle$ and $|-\rangle$ as

$$|0\rangle \to |000\rangle, \quad |1\rangle = |111\rangle,$$

thus obtaining the logical qubit state vectors as

$$|0_L\rangle = \frac{(|000\rangle + |111\rangle)(|000\rangle + |111\rangle)(|000\rangle + |111\rangle)}{\sqrt{8}}, \tag{10.23}$$

$$|1_L\rangle = \frac{(|000\rangle - |111\rangle)(|000\rangle - |111\rangle)(|000\rangle - |111\rangle)}{\sqrt{8}}. \tag{10.24}$$

The error syndrome is obtained by measuring the following operators:

$$(\sigma_1^X\sigma_2^X\sigma_3^X)(\sigma_4^X\sigma_5^X\sigma_6^X); \qquad (\sigma_4^X\sigma_5^X\sigma_6^X)(\sigma_7^X\sigma_8^X\sigma_9^X); \tag{10.25}$$

$$\sigma_1^Z\sigma_2^Z; \quad \sigma_2^Z\sigma_3^Z; \quad \sigma_4^Z\sigma_5^Z; \quad \sigma_5^Z\sigma_6^Z; \quad \sigma_7^Z\sigma_8^Z; \quad \sigma_8^Z\sigma_9^Z. \tag{10.26}$$

Note that the operators (10.25) deserve to detect phase-flip errors (which change sign on a block of three qubits), while (10.26) to detect bit-flip errors. Then, a single qubit error can be undone by applying the unitary σ^X or σ^Z or $\sigma^X\sigma^Z$ on the qubit on which the error was located.

One could say that the correction of bit-, phase-, and bit-phase-flip errors is not enough because there exists a continuum of errors than cannot be corrected. Remarkably, it is possible to show that the correction of these 3 errors is sufficient to correct any kind of single qubit error.

To show that, let us consider a generic qubit CPTP map $\mathcal{N}$ with Kraus decomposition

$$\mathcal{N}(\rho) = \sum_j A_j \rho A_j^\dagger.$$

Since the Kraus operators are generic operators on $\mathbb{C}^2$, they can hence be expanded in terms of Pauli operators — including the identity I — as

$$A_j = a_{j0}\, I + a_{j1}\, \sigma^X + a_{j2}\, \sigma^Y + a_{j3}\, \sigma^Z,$$

with coefficients $a_{jk} \in \mathbb{C}$.

Then, given an input state vector $|\psi\rangle$, if the error A_j occurred, the state is transformed to

$$A_j|\psi\rangle = a_{j0} I|\psi\rangle + a_{j1}\, \sigma^X |\psi\rangle + a_{j2}\, \sigma^Y |\psi\rangle + a_{j3}\, \sigma^Z |\psi\rangle.$$

However, detecting a Pauli error is a measurement process. Hence, if for instance the error σ^X is detected, the state will collapse to $\sigma^X |\psi\rangle$. Then, the error will be undone by applying the unitary σ^X, that will recover the original state vector $|\psi\rangle$.

10.4 Basics of quantum error correction theory

We can now provide a general formulation of quantum error correcting theory.

Definition 10.4.1. A quantum code $\mathcal{C}$ is a subspace of $(\mathbb{C}^2)^{\otimes n}$ of dimension 2^k (with $k \leq n$), able to encode k qubit into n.

As a consequence, any linear combination of the basis codewords spanning $\mathcal{C}$ is also a valid codeword, corresponding to the same linear combination of the unencoded basis states.

An error operator E can be considered as an element of the set

$$\mathrm{P}(n) := \{\pm I, \pm iI, \pm\sigma^X, \pm i\sigma^X, \pm\sigma^Y, \pm i\sigma^Y, \pm\sigma^Z, \pm i\sigma^Z\}^{\otimes n}, \quad (10.27)$$

which takes the structure of a group under multiplication and is known as the *Pauli group.* Note that the factors $\pm$ and $\pm i$ are global phases and have thus no relevance for the action and correction of the errors.

Given a quantum channel $\mathcal{N} \leftrightarrow \{E_a\}$, the fidelity of the code is determined by the composition of the recovery map $\mathcal{R} \leftrightarrow \{R_r\}$ with the channel, $\mathcal{R} \circ \mathcal{N}$, restricted to $\mathcal{C}$ and results as

$$F = \min_{|\Psi\rangle \in \mathcal{C}} \sum_{r,a} |\langle \Psi | R_r E_a | \Psi \rangle|^2 . \tag{10.28}$$

As a consequence, the error of the code reads

$$\max_{|\Psi\rangle \in \mathcal{C}} \sum_{r,a} \| (R_r E_a - \langle \Psi | R_r E_a | \Psi \rangle) |\Psi\rangle \|^2 . \tag{10.29}$$

Thus, to correct a set $\mathcal{E} \subset \mathrm{P}(n)$ of errors, it is required that

$$\| (R_r E_a - \langle \Psi | R_r E_a | \Psi \rangle) |\Psi\rangle \| = 0, \quad \forall\, |\Psi\rangle \in \mathcal{C},\ E_a \in \mathcal{E}. \tag{10.30}$$

This implies that $R_r E_a |\Psi\rangle = \lambda_{ra}(|\Psi\rangle)|\Psi\rangle$. By linearity of $R_r E_a$, it results $\lambda_{ra}(|\Psi\rangle) = \lambda_{ra}$, i.e. independent of $|\Psi\rangle$.

Necessary and sufficient condition for the code $\mathcal{C}$ to correct a set of errors $\mathcal{E} \subset \mathrm{P}(n)$ are established by the following theorem.

Theorem 10.4.2. *Given a code $\mathcal{C} \subseteq (\mathbb{C}^2)^{\otimes n}$, the necessary and sufficient condition to correct a set of errors $\mathcal{E}$ is that*

$$\langle i_L | E_a^\dagger E_b | j_L \rangle = \mathsf{C}_{ab} \delta_{ij}, \quad E_a, E_b \in \mathcal{E}, \tag{10.31}$$

for some (Hermitian) matrix with entries C_{ab}.

Proof. To prove the necessity of the condition (10.31), we assume the correctability of $\mathcal{E}$, i.e. the existence of a recovery map $\mathcal{R} \leftrightarrow \{R_r\}$. Thus, it will be $R_r E_a |\Psi\rangle = \lambda_{ra} |\Psi\rangle$ for $|\Psi\rangle \in \mathcal{C}$. Then

$$\langle i_L | E_a^\dagger E_b | j_L \rangle = \sum_r \langle i_L | E_a^\dagger R_r^\dagger R_r E_b | j_L \rangle = \sum_r \lambda_{ar}^* \lambda_{rb} \langle i_L | j_L \rangle = \mathsf{C}_{ab} \delta_{ij},$$

where $\mathsf{C}_{ab} := \sum_r \lambda_{ar}^* \lambda_{rb}$.

Let us now prove that the condition (10.31) is also sufficient for the correctability of $\mathcal{E}$. The matrix C is Hermitian, so it can be diagonalized with a suitable unitary U, i.e. $\sum_{ab} \mathsf{U}_{ha} \mathsf{C}_{ab} \mathsf{U}_{kb}^* = \delta_{hk} d_k$. Then, by introducing the error operators $F_h^\dagger := \frac{1}{\sqrt{d_h}} \sum_a \mathsf{U}_{ha} E_a^\dagger$, we get either

$$\langle i_L | F_a^\dagger F_b | j_L \rangle = \delta_{ab}\delta_{ij}, \tag{10.32}$$

or

$$\langle i_L | F_a^\dagger F_b | j_L \rangle = 0, \tag{10.33}$$

depending on whether all $d_k \neq 0$, or not, i.e. on whether C is non-singular or singular. The latter case implies the presence of some errors that annihilate any codeword, so the probability of one occurring is strictly zero and we need not consider them. The other errors always produce orthogonal states, so we can make some measurement that will tell us exactly which error occurred, and at this point it is a simple matter to correct it. □

If we refer to Shor's nine-qubit code of Section 10.3, we can see that σ_1^Z and σ_2^Z act the same way on the code, so $\sigma_1^Z - \sigma_2^Z$ will annihilate codewords. This phenomenon will occur if and only if C_{ab} does not have maximum rank. A code for which C_{ab} is singular is called a *degenerate code*, while a code for which it is not is *non-degenerate.* Shor's nine-qubit code is clearly degenerate.

It is worth noting that in contrast to the classical codes, for the quantum codes two erroneous versions of the same codeword do not need to be orthogonal (perfectly distinguishable), however it must be $\langle i_L | E_a^\dagger E_b | i_L \rangle = \langle j_L | E_a^\dagger E_b | j_L \rangle$ for all basis codewords.

Concerning the recovery map $\mathcal{R}$, in practice it is realized by Kraus operators

$$R_l = \sum_i |i_L\rangle\langle v_l^{i_L}| \tag{10.34}$$

where $|v_l^{i_L}\rangle \propto E_l|i_L\rangle$ (the proportionality symbol means that $E_l|i_L\rangle$ must be normalized). In other words, it is $R_l \propto \sum_i |i_L\rangle\langle i_L|E_l^\dagger$, which means that R_l is build up with the inverse of E_l followed by the projector onto the code space.

Finally, the weight t of $E \in \mathrm{P}(n)$ is the number of single-qubit components which are different from identity. In Theorem 10.4.2, $E_a^\dagger E_b$ is still in the group $\mathrm{P}(n)$ when E_a and E_b are in $\mathrm{P}(n)$.

The weight of the smallest E in $\mathrm{P}(n)$ for which (10.31) does not hold is called the distance of the code. A quantum code to correct up to t errors must have distance at least $2t+1$. A minimum distance d code encoding k qubits in n qubits is denoted as an $[[n,k,d]]$ -code.[3]

10.4.1 *Bounds on quantum codes*

For a non-degenerate code with basis codewords $|i_L\rangle$ and possible errors E_a, all of the states $E_a|i_L\rangle$ are linearly independent for all a and i. If the code uses n qubits, there can only be $2n$ linearly indepedent vectors in the Hilbert space, so the number of errors times the number of codewords must be less than or equal to $2n$. If the code corrects all errors of weight t or less and encodes k qubits, i.e it is a $[[n.k,d]]$ code, this means

$$2^k \sum_{j=0}^{t} 3^i \binom{n}{j} \leq 2^n. \tag{10.35}$$

There are $\binom{n}{j}$ ways to choose j qubits to be affected by j errors and 3^j ways these errors can be tensor products of σ^X, $\sigma^Y = i\sigma^X\sigma^Z$, and σ^Z. This is the *quantum Hamming bound.*

For large n and t/n constant, this becomes (using the Stirling approximation)

$$\frac{k}{n} \lesssim 1 - H_2(t/n) - \frac{t}{n}\log 3. \tag{10.36}$$

For degenerate codes, we can still set a bound, but it will not be as restrictive. For an $[[n,k,d]]$ code, we can choose any $d-1$ qubits and remove them. The remaining $n-d+1$ qubits must contain enough information to reconstruct not only the 2^k possible codewords, but the state of the missing qubits as well. Because the missing qubits can be any qubits, we can choose them to have maximum entropy.

[3]For a generic quantum code without specified minimum, we will use the notation $[[n,k]]$.

Then

$$n - d + 1 \geq d - 1 + k \tag{10.37}$$
$$n \geq 2(d-1) + k. \tag{10.38}$$

This bound is a quantum analog of the classical Singleton bound. A code to correct t errors must have distance $d = 2t + 1$, so for such a code, $n \geq 4t + k$. Actually, this bound holds true for any code with a given minimum distance, whether it is degenerate or non-degenerate. It is worth remarking that this bound demonstrates that the smallest one-error correcting quantum code uses five qubits. Actually, a $[[5, 1, 3]]$ code exists as we will see.

We can also set a lower bound on the rate of quantum codes. Recall the condition (10.31) for a quantum code correcting errors $\{E_a\}_a$ with basis codewords $\{|i_L\rangle\}_i$. The matrix C_{ab} is Hermitian, but is further constrained by the algebraic relationships of the operators $E_a^\dagger E_b$. So, it is better to consider C_{ab} as a function of operators $O = E_a^\dagger E_b$. When the possible errors are all operators of up to weight t, O can be any operator of weight $\leq 2t$. More generally, for a code of distance d, O is any operator of weight less than d. Therefore, the statement

$$\langle i_L | E_a^\dagger E_b | i_L \rangle = \mathsf{C}_{ab}, \tag{10.39}$$

is actually

$$N = \sum_{j=0}^{d-1} 3^j \binom{n}{j}$$

constraints on the state vector $|i_L\rangle$. For generic C_{ab} (satisfying the appropriate constraints) and generic linear subspace V with dimension larger than N, there will be state vectors $|\psi\rangle$ satisfying (10.39).

Suppose we choose generic C_{ab} and a generic state vector $|\psi_1\rangle$ satisfying (10.39). Now focus our attention on the subspace orthogonal to $|\psi_1\rangle$ and to all $O|\psi_1\rangle$ for operators O of weight less than d. For an n-qubit Hilbert space, this subspace has dimension $2n - N$. Then, choose a generic state vector $|\psi_2\rangle$ in this subspace satisfying (10.39). Restrict further the attention to the subspace orthogonal to both $O|\psi_1\rangle$ and $O|\psi_2\rangle$. We can again pick $|\psi_3\rangle$ in this subspace satisfying (10.39), and so on. Choose $|\psi_i\rangle$ orthogonal to all $O|\psi_j\rangle$ ($j \leq i-1$) and satisfying (10.39). We can continue doing this until

$$i \sum_{j=0}^{d-1} 3^j \binom{n}{j} \geq 2^n.$$

Therefore, we can always find a distance d quantum code encoding k qubits in n qubits satisfying the *quantum Gilbert–Varshamov bound*

$$k \geq \max \left\{ k' \middle| 2^{k'} \sum_{j=0}^{2t} 3^j \binom{n}{j} \leq 2^n \right\}. \tag{10.40}$$

For large n and t/n fixed, it yields

$$\frac{k}{n} \gtrsim 1 - H_2(2t/n) - \frac{2t}{n} \log 3. \tag{10.41}$$

10.5 Stabilizer codes

Unfortunately, the explicit construction of quantum codes is not an easy task. Here, we provide a recipe that relies on classical linear codes.

Let us begin to note some useful properties of the Pauli group: (i) every element of $\mathrm{P}(n)$ squares to $\pm I_n$; (ii) any two elements of $\mathrm{P}(n)$ either commute or anti-commute; (iii) every element of $\mathrm{P}(n)$ is unitary; (iv) elements of $\mathrm{P}(n)$ are either Hermitian or anti-Hermitian.

Definition 10.5.1. A stabilizer code $\mathcal{C} \subseteq (\mathbb{C}^2)^{\otimes n}$ is the set of vectors stabilized by an abelian subgroup $\mathrm{S} \subseteq \mathrm{P}(n)$, i.e.

$$\mathcal{C} = \left\{ |x\rangle \in \mathbb{C}^{2\otimes n} \,\middle|\, M|x\rangle = |x\rangle, \forall M \in \mathrm{S} \right\}.$$

We should note that for group S to be the stabilizer of a non-trivial subspace, it must satisfy two conditions: the elements of S commute, and $-I_n$ is not in S (this latter condition implies that all elements of S are Hermitian, and hence have eigenvalues ± 1.)

For an $[[n,k]]$ stabilizer code, which encodes k logical qubits into n physical qubits, $\mathcal{C}$ has dimension 2^k and S has 2^{n-k} elements. Being abelian, S can be specified by a set of $n-k$ generators, $\{M_i\}_{i=1}^{n-k}$.

Suppose $\mathcal{C}(\mathrm{S})$ is a stabilizer code, and the quantum register is subject to errors from a set $\mathcal{E} = \{E_a\} \subset \mathrm{P}(n)$. How are the error correcting properties of $\mathcal{C}(\mathrm{S})$ related to the generators of S?

First, suppose that E_a anti-commutes with a particular stabilizer generator M_i of S. Then

$$M_i E_a |\psi\rangle = -E_a M_i |\psi\rangle = -E_a |\psi\rangle.$$

$E_a|\psi\rangle$ is an eigenvector of M_i with eigenvalue -1, and hence must be orthogonal to the code space (all of whose vectors have eigenvalue $+1$). As the error operator E_a takes the code space of $\mathcal{C}(\mathrm{S})$ to an orthogonal subspace, an occurrence of E_a can be detected by measuring M_i.

What if E_a commutes with the generators of S? If $E_a \in \mathrm{S}$, we do not need to worry, because the error does not corrupt the space at all. The real danger comes when E_a commutes with all the elements of S, but is not itself in S. The set of elements in $\mathrm{P}(n)$ that commute with all of S is the centralizer $\mathrm{C_P(S)}$ of S. If $E \in \mathrm{C_P(S)} - \mathrm{S}$, then E changes elements of $\mathcal{C}(\mathrm{S})$, but does not take them out of $\mathcal{C}(\mathrm{S})$. Thus, if $M \in \mathrm{S}$ and $|\psi\rangle \in \mathcal{C}(\mathrm{S})$, then

$$ME|\psi\rangle = EM|\psi\rangle = E|\psi\rangle.$$

Because $E \notin \mathrm{S}$, there is some state of $\mathcal{C}(\mathrm{S})$ that is not fixed by E. E will be an undetectable error for this code.

Putting the above cases together, and referring to Theorem 10.4.2, we can conclude that a stabilizer code $\mathcal{C}(\mathrm{S})$ can correct a set of errors $\mathcal{E} \subset \mathrm{P}(n)$ if and only if $E_a^\dagger E_b \in \mathrm{S} \cup (\mathrm{P}(n) - \mathrm{C_P(S)})$ for all $E_a, E_b \in \mathcal{E}$.

Example. Shor's code considered as a stabilizer code is generated by

$$\begin{array}{l} M_1 = \sigma^Z \ \sigma^Z \ I \ I \ I \ I \ I \ I \ I \\ M_2 = \sigma^Z \ I \ \sigma^Z \ I \ I \ I \ I \ I \ I \\ M_3 = I \ I \ I \ \sigma^Z \ \sigma^Z \ I \ I \ I \ I \\ M_4 = I \ I \ I \ \sigma^Z \ I \ \sigma^Z \ I \ I \ I \\ M_5 = I \ I \ I \ I \ I \ I \ \sigma^Z \ \sigma^Z \ I \\ M_6 = I \ I \ I \ I \ I \ I \ \sigma^Z \ I \ \sigma^Z \\ M_7 = \sigma^X \ \sigma^X \ \sigma^X \ \sigma^X \ \sigma^X \ \sigma^X \ I \ I \ I \\ M_8 = \sigma^X \ \sigma^X \ \sigma^X \ I \ I \ I \ \sigma^X \ \sigma^X \ \sigma^X \end{array}.$$

Nevertheless, the correction ability of a stabilizer code can be better studied as follows.

First, note that any error operator belonging to $\mathrm{P}(n)$ can be written as $E = i^{\xi} \, \overrightarrow{\sigma_x^X} \overrightarrow{\sigma_z^Z}$ where $\xi = 0, 1, 2, 3$, $x, z \in \mathbb{F}_2^n$ and e.g.

$$\overrightarrow{\sigma_x^X} := I_1 \otimes \sigma_2^X \otimes \cdots \otimes I_n, \quad x = 01 \cdots 0,$$

$$\overrightarrow{\sigma_z^Z} := \sigma_1^Z \otimes I_2 \otimes \cdots \otimes \sigma_n^Z, \quad z = 10 \cdots 1.$$

Then, the elements of $\mathrm{P}(n)$, getting rid of the $\pm$ and $\pm i$ in front of them since they are irrelevant for error correction (i.e. considering the quotient group $\mathrm{P}(n)/\{\pm I, \pm iI\}$), can be related to $2n$-dimensional binary vectors

$$E \in \mathrm{P}(n)/\{\pm I, \pm iI\} \leftrightarrow (x|z) \equiv e \in \mathbb{F}_2^n \times \mathbb{F}_2^n. \tag{10.42}$$

Then, by using the vectors $m_1, m_2, \ldots, m_{n-k}$ in $\mathbb{F}_2^n \times \mathbb{F}_2^n$ corresponding to the generators $M_1, M_2, \ldots, M_{n-k}$ of S, it is possible to write down the following $(n-k) \times 2n$ matrix

$$\mathsf{H} := \begin{pmatrix} m_{1,x}|m_{1,z} \\ \vdots \\ m_{n-k,x}|m_{n-k,z} \end{pmatrix}. \tag{10.43}$$

Now, the commutativity of the generators M_i translates into the following condition:

$$m_{i,x} \cdot m_{j,z} + m_{j,x} \cdot m_{i,z} = 0, \quad \forall i, j = 1, \ldots, n-k, \tag{10.44}$$

for vectors in $\mathbb{F}_2^n$, or in other words $\mathsf{H}_x\mathsf{H}_z^\top + \mathsf{H}_z\mathsf{H}_x^\top = 0$, once we write $\mathsf{H} \equiv (\mathsf{H}_x|\mathsf{H}_z)$. In turn, this condition guarantees the linear independence of the rows of H in (10.43), which can then be considered as a parity-check matrix of a $[2n, k]$ classical linear code Γ corresponding to $\mathcal{C}$.

Then the analysis of the correction ability of the $[[n, k]]$ quantum code $\mathcal{C}$ can be traced back to that of corresponding classical $[2n, k]$ linear code $\mathcal{C}$.

Actually, a set of errors $\{E_i\}_i \subseteq \mathrm{P}(n)$ can be corrected if and only if for every $(e_{i,x}|e_{i,z})$, $(e_{j,x}|e_{j,z})$ it is

$$\mathsf{H}(e_{i,x} + e_{j,x}|e_{i,z} + e_{j,z})^\top \neq 0, \tag{10.45}$$

or in other words, if the syndrome of e_i differs from the syndrome of e_j.

Note that to extract the syndrome is enough to measure the stabilizer generators $M_1 M_2 \cdots M_{n-k}$.

A stabilizer code of minimum distance d has the property that each element of $\mathrm{P}(n)$ of weight t less than $d/2$ either lies in the stabilizer or anti-commutes with some element of the stabilizer.

Example. The $[[5, 1, 3]]$ that saturates the Hamming bound is generated by

$$\begin{array}{l} M_1 = \sigma^X\ \sigma^Z\ \sigma^Z\ \sigma^X\ I \\ M_2 = \ I\ \sigma^X\ \sigma^Z\ \sigma^Z\ \sigma^X \\ M_3 = \sigma^X\ I\ \sigma^X\ \sigma^Z\ \sigma^Z \\ M_4 = \sigma^Z\ \sigma^X\ I\ \sigma^X\ \sigma^Z \end{array}.$$

If we consider a classical linear code Γ with parity-check matrix H_Γ and build up the parity-check matrix of (10.43) as

$$\mathsf{H} = \begin{pmatrix} \mathsf{H}_\Gamma & 0 \\ 0 & \mathsf{H}_\Gamma \end{pmatrix}, \tag{10.46}$$

then the condition (10.44) reads $\mathsf{H}_\Gamma\mathsf{H}_\Gamma^\top = 0$. Remembering that $\mathsf{H}_\Gamma = \mathsf{G}_{\Gamma^\perp}$, this condition will be satisfied if it happens that $\Gamma^\perp \subseteq \Gamma$. Therefore, in such a case we get an $[[n, 2k - n, d]]$ quantum code from an $[n, k, d]$ classical one.

Example. The $[[7, 1, 3]]$ coming from the $[7, 4, 3]$ Hamming code having parity-check matrix H_Γ equal to (10.9).

Exercises

1. Show that the set $\mathcal{E}_n$ of all even-weight vectors of the Hamming space $\mathbb{F}_2^n$ is a subspace (what is the dimension? write down a basis). Further, prove that $\mathcal{E}_n^{\perp}$ is a repetition code of length n.
2. Assume that on three qubits only the following errors can occur:
$$\{III, I\sigma^X\sigma^X, I\sigma^Z\sigma^Z, I(\sigma^Z\sigma^X)(\sigma^Z\sigma^X)\}$$
Describe how an arbitrary error
$$E = \alpha III + \beta I\sigma^X\sigma^X + \gamma I\sigma^Z\sigma^Z + \delta I(\sigma^Z\sigma^X)(\sigma^Z\sigma^X),$$
where $\alpha, \beta, \gamma, \delta \in \mathbb{C}$, on the state
$$\frac{1}{\sqrt{2}}(|00\rangle + |11\rangle)|\psi\rangle$$
can be corrected to obtain $|\psi\rangle$.
3. Construct stabilizer generators for an $[[n,k,d]] = [[3,2,0]]$ quantum code (able to detect any single-qubit error) and find the encoded state. Are all $[[3,2,0]]$ stabilizer codes equivalent? [Two quantum codes $\mathcal{C}_1$ and $\mathcal{C}_2$ are equivalent if a permutation of qubits, together with single-qubit unitaries, transforms the subspace of $\mathcal{C}_1$ into that of $\mathcal{C}_2$.]
4. Suppose Γ_1 and Γ_2 are $[n, k_1]$ and $[n, k_2]$ classical linear codes such that $\Gamma_2 \subset \Gamma_1$. Furthermore, let Γ_1 and $\Gamma_2^{\perp}$ be codes both correcting t errors. Show that the quantum code with codewords $|x + \Gamma_2\rangle := \frac{1}{\sqrt{|\Gamma_2|}} \sum_{y \in \Gamma_2} |x + y\rangle$, where $x \in \Gamma_1$, is a $[[n, k_1 - k_2]]$ quantum code able to correct all errors on t qubits.
5. Given a quantum channel $\mathcal{N}$ with Kraus operators $\{A_i\}_i$, define $\mathrm{rank}(\mathcal{N}) := \mathrm{rank}(\chi)$, where $\mathcal{N}(\rho) = \sum_i A_i \rho A_i^{\dagger} = \sum_{m,n} \chi_{m,n} B_m \rho B_n^{\dagger}$ with B_is satisfying the orthogonality condition $\mathrm{Tr}(B_i^{\dagger} B_j) = d\delta_{i,j}$, d being the dimension of the system. Then show that for any non-degenerate code subspace $\mathcal{C}$, it must be
$$d \geq |\mathcal{C}|\, \mathrm{rank}(\mathcal{N}).$$
Also, recover the usual Hamming bound for noise acting independently on the encoding systems.

Further, consider the following quantum channel:

$$\mathcal{N}(\rho) = p\rho + \sum_{i=1\ldots n, j>i} (p_{X,ij}\sigma_i^X \sigma_j^X \rho \sigma_j^X \sigma_i^X + p_{Y,ij}\sigma_i^Y \sigma_j^Y \rho \sigma_j^Y \sigma_i^Y$$
$$+ p_{Z,ij}\sigma_i^Z \sigma_j^Z \rho \sigma_j^Z \sigma_i^Z),$$

where with probability $p = 1 - \sum_{i=1\ldots n, j>i} p_{X,ij} + p_{Y,ij} + p_{Z,ij}$ the input state is left unchanged, while with probabilities $p_{X,ij}$, $p_{Y,ij}$ and $p_{Z,ij}$ pairwise Pauli operators σ^X, σ^Y and σ^Z are applied to qubits i and j, respectively. Take $k = 1$ giving $n = 7$ and show that if errors are pairwise completely correlated, it is possible to exploit degeneracy to violate the bound and achieve perfect quantum error correction.

Chapter 11

CHANNEL CAPACITIES

By means of error correcting codes, we have seen the possibility of sending information reliably through noisy channels. However, the rate at which this can be done strongly depends on the employed code. Being interested in the maximum of such rates, we shall give here a rigorous definition of channel capacities and show by coding theorems that the upper bounds on transmission rates following from Chapter 9 are actually achievable.

11.1 Shannon noisy coding theorem

Let us consider a classical *memoryless* channel $N : \mathfrak{P}(\mathcal{X}) \to \mathfrak{P}(\mathcal{Y})$, i.e. a stochastic map that acts independently and identically on each input symbol.

Given a set of message $\mathcal{M}$, a code is specified by encoding and decoding maps

$$E : \mathcal{M} \to \mathcal{X}^n \tag{11.1}$$

$$\Delta : \mathcal{Y}^n \to \mathcal{M}. \tag{11.2}$$

It has (average) error probability

$$p_e = \frac{1}{M} \sum_{m=1}^{M} \Pr\{\Delta \circ N^n \circ E(m) \neq m\} \tag{11.3}$$

and rate $\frac{1}{n} \log |\mathcal{M}|$.

Definition 11.1.1. A communication rate $\mathtt{R}$ is said to be achievable if there exists a sequence of encoding/decoding maps such that

$$\lim_{n\to\infty} \frac{1}{n} \log |\mathcal{M}| = \mathtt{R}, \quad \text{and} \quad \lim_{n\to\infty} p_e = 0.$$

The supremum among achievable communication rates defines the capacity of the communication channel, $C(N)$.

Shannon's second theorem gives a closed expression for the capacity of communication channel.

Theorem 11.1.2 (Shannon noisy channel coding theorem). *The capacity of the communication channel $X \xrightarrow{N} Y$, with X and Y input and output random variables in their respective alphabets, is given by*

$$C(N) = \max_X I(X:Y).$$

Before giving the proof of the theorem, we need to extend the notions about typical sequences introduced in Section 8.1.1.

Definition 11.1.3. The set ${}_JT^{(n)}$ of jointly typical sequences (x^n, y^n) with respect to the probability distribution $p(x^n, y^n)$ is the set of sequences with empirical entropies ϵ-close to the true entropies, i.e.

$$\left| -\frac{1}{n} \log p(x^n) - H(X) \right| < \epsilon, \quad \left| -\frac{1}{n} \log p(y^n) - H(Y) \right| < \epsilon$$

and

$$\left| -\frac{1}{n} \log p(x^n, y^n) - H(X,Y) \right| < \epsilon$$

where $p(x^n, y^n) = \prod_{i=1}^n p(x_i, y_i)$.

Properties:

- $\Pr((X^n, Y^n) \in {}_JT^{(n)}) \to 1$ for $n \to \infty$.
- $|{}_JT^{(n)}| \leq 2^{n(H(X,Y)+\epsilon)}$.

- If X^n and Y^n are drawn according to $p(x^n)$ and $p(y^n)$, i.e. the marginals of $p(x^n, y^n)$, then

$$(1-\epsilon)2^{-n(I(X:Y)+3\epsilon)} \leq \Pr((X^n, Y^n) \in {}_J T^{(n)}) \leq 2^{-n(I(X:Y)-3\epsilon)}$$

for sufficiently large n.

Definition 11.1.4. A sequence y^n is conditionally typical to the sequence x^n if the latter is typical and (x^n, y^n) is jointly typical.

Properties:

- If y^n is conditionally typical to x^n, then

$$2^{-n(H(X,Y)-H(X)+2\delta)} \leq p_{y^n|x^n} \leq 2^{-n(H(X,Y)-H(X)-2\delta)}.$$

- Consider the pairs (x^n, y^n) for which y^n is conditionally typical to x^n, and denote the sum over such y^n by a prime, then

$$\sum_{x^n} p_{x^n} \sum_{y^n|x^n}{}' p_{y^n} > 1 - 2\epsilon. \tag{11.4}$$

Proof of Theorem 11.1.2. *Converse part.* This part aims at proving that the rate $\mathtt{R}$ is upper bounded by $\max_X H(X:Y)$. By referring to the classical Fano inequality 9.1.1, we have for n channel uses

$$H(X^n|Y^n) \leq H_2(p_e) + p_e \log(|\mathcal{M}| - 1) \leq H_2(p_e) + p_e n\mathtt{R}.$$

This in turn implies

$$I(X^n : Y^n) \geq H(X^n) - H_2(p_e) - p_e n\mathtt{R}.$$

Assuming the input messages are uniformly distributed, it results $H(X^n) = \log|\mathcal{M}| = n\mathtt{R}$. Thus, we obtain

$$\mathtt{R} \leq \frac{H_2(p_e)}{n(1-p_e)} + \frac{I(X^n : Y^n)}{n(1-p_e)}. \tag{11.5}$$

Now we may note that

$$I(X^n : Y^n) = H(Y^n) - H(Y^n|X^n) \tag{11.6}$$

$$= H(Y^n) - \sum_{i=1}^{n} H(Y_i|Y_1, \ldots, Y_{i-1}, X^n) \tag{11.7}$$

$$= H(Y^n) - \sum_{i=1}^{n} H(Y_i|Y_i) \tag{11.8}$$

$$\leq \sum_{i=1}^{n} H(Y_i) - \sum_{i=1}^{n} H(Y_i|Y_i), \tag{11.9}$$

where (11.7) is obtained by chain rule [property (iv) of Section 2.5], (11.8) by the fact that Y_i only depends on X_i (the channel is memoryless and i.i.d.) and (11.9) by subadditivity of Shannon entropy [property (i) of Section 2.5].

Plugging this into (11.5) gives

$$\mathtt{R} \leq \frac{H_2(p_e)}{n(1-p_e)} + \frac{\sum_{i=1}^{n} I(X_i : Y_i)}{n(1-p_e)}. \tag{11.10}$$

Finally, the limit $p_e \to 0$ yields $\mathtt{R} \leq \max_X I(X : Y)$.

Direct part. This part aims at proving that the rate $\max_X I(X:Y)$ is actually achievable. To this end, we generate an $[n\mathtt{R}, n]$ code at random according to the distribution $p(x)$. Specifically, we independently generate $2^{n\mathtt{R}}$ codewords (one for each message $m \in \mathcal{M}$) according to the distribution

$$p(x^n) = \prod_{i=1}^{n} p(x_i), \tag{11.11}$$

and we arrange such codewords as rows of a matrix

$$\Gamma = \begin{pmatrix} x_1(1) & x_2(1) & \ldots & x_n(1) \\ \vdots & \vdots & \ddots & \vdots \\ x_1(2^{n\mathtt{R}}) & x_2(2^{n\mathtt{R}}) & \ldots & x_n(2^{n\mathtt{R}}) \end{pmatrix}. \tag{11.12}$$

Each entry in this matrix is generated i.i.d. according to $p(x)$, hence the probability to generate a particular code Γ is

$$\Pr(\Gamma) = \prod_{m=1}^{2^{nR}} \prod_{i=1}^{n} p(x_i(m)). \tag{11.13}$$

A message M (intended as a random variable) is chosen according to the uniform distribution $\Pr(m) = 2^{-nR}$ (with $m = 1, 2, \ldots, 2^{nR}$). The mth codeword $X^n(m)$ corresponding to the mth row of Γ is sent over the channel. The receiver receives a sequence Y^n according to the distribution

$$p(y^n|x^n(m)) = \prod_{i=1}^{n} p(y_i|x_i(m)), \tag{11.14}$$

and guesses which message was sent. The receiver declares that the index $\tilde{M}$ was sent if (i) $(X^n(\tilde{M}), Y^n)$ is jointly typical; (ii) there is no other index k such that $(X^n(k), Y^n) \in {}_J T^{(n)}$. There is a decoding error if $\tilde{M} \neq M$.

Let us then evaluate the average (overall possible codes) probability of error

$$p_e = \frac{1}{2^{nR}} \sum_{m=1}^{2^{nR}} \sum_{\Gamma} p(\Gamma) p_e(m|\Gamma), \tag{11.15}$$

where $p_e(m|\Gamma)$ is the probability of error for the message m given the code Γ. By the symmetry of the code, the quantity $\sum_\Gamma p(\Gamma) p_e(m|\Gamma)$ will not depend on the particular index m that was sent. Thus, without loss of generality we can assume that the message $M = 1$ was sent so that it results $\Pr_e = \Pr(\text{error}|M = 1)$.

We now define the events Z_i by which the ith codeword and Y^n are jointly typical

$$Z_i := \{(X^n(i), Y^n) \in {}_J T^{(n)}\}, \quad i = 1, 2, \ldots, 2^{nR}. \tag{11.16}$$

Recall that Y^n is the result of sending the first codeword $X^n(1)$ over the channel. Then, an error occurs in the decoding scheme if either Z_1^c occurs (the complementary event to Z_1, i.e. the transmitted codeword and the received sequence are not jointly typical) or $Z_2 \cup Z_3 \cup \ldots \cup Z_{2^{nR}}$ occurs (a wrong codeword is jointly typical with the received sequence). Hence, we have

$$p_e = \Pr(\text{error}|M = 1) = p(Z_1^c \cup Z_2 \cup Z_3 \cup \ldots \cup Z_{2^{n\mathtt{R}}})$$
$$\leq p(Z_1^c) + \sum_{i=2}^{2^{n\mathtt{R}}} p(Z_i). \tag{11.17}$$

Now by the joint typicality property, $p(Z_1^c) \leq \epsilon$ for n sufficiently large. Since by the way the code is generated $X^n(1)$ and $X^n(i)$ are independent, so are Y^n and $X^n(i)$ for $i \neq 1$. Thus, the probability that $X^n(i)$ and Y^n are jointly typical is $\leq 2^{-n(I(X:Y)-3\epsilon)}$. Consequently,

$$p_e = \Pr(\text{error}|M = 1) \leq p(Z_1^c) + \sum_{i=2}^{2^{n\mathtt{R}}} p(Z_i)$$
$$\leq \epsilon + \sum_{i=2}^{2^{n\mathtt{R}}} 2^{-n(I(X:Y)-3\epsilon)}$$
$$\leq 2\epsilon, \tag{11.18}$$

if n is sufficiently large and $\mathtt{R} < I(X : Y) - 3\epsilon$. □

Remark. Random coding is often used (as we shall see) to prove coding theorems, however, it is not a practical coding scheme to be used. Codes are selected at random in the proof merely to symmetrize the mathematics and to show the existence of a good code. However, without any structure in the coding procedure, the decoding will become extremely difficult (the simple scheme of table lookup would require an exponentially large table).

Example. The capacity of the binary symmetric channel with error probability q. Let p (resp., $(1 - p)$) be the probability of sending "0" (resp., "1") in input to the channel.

$$C = \max_p [H(Y) - H(Y|X)]$$
$$= \max_p [H(Y) - \sum_x p(x) H(Y|X = x)]$$
$$= \max_p [H(Y) - H_2(q)] = 1 - H_2(q),$$

achieved for $p = 1/2$.

11.2 Holevo–Schumacher–Westmoreland (HSW) coding theorem

Consider now Alice sending classical information encoded into quantum states through a *memoryless* quantum channel $\mathcal{N} : \mathfrak{D}(\mathcal{H}_A) \to \mathfrak{D}(\mathcal{H}_B)$, i.e. a CPTP map that acts independently and identically on each input system.

The code is specified by considering a set $\mathcal{M}$ of possible messages with associated Hilbert space $\mathcal{H}_{\mathcal{M}} = \text{span}\{|m\rangle\}_{m\in\mathcal{M}}$, an encoding map

$$\mathcal{E} : \mathfrak{D}(\mathcal{H}_{\mathcal{M}}) \to \mathfrak{D}(\mathcal{H}_A^{\otimes n}), \tag{11.19}$$

that associates to a message $m \in \mathcal{M}$ (i.e. to $|m\rangle \in \mathcal{H}_{\mathcal{M}}$) an n-system *product* state

$$\rho_m \equiv \rho_{x^n(m)} = \rho_{x_1} \otimes \rho_{x_2} \otimes \cdots \otimes \rho_{x_n},$$

where $x^n(m)$ is a codeword of length n corresponding to the message m. The output state will consequently be

$$\mathcal{N}(\rho_{x_1}) \otimes \mathcal{N}(\rho_{x_2}) \otimes \cdots \mathcal{N}(\rho_{x_n}).$$

On this output, Bob will perform a measurement by POVM elements $\{\Xi_{y^n}\}_{y^n}$ where Ξ_{y^n} is an operator in the Hilbert space $\mathcal{H}^{\otimes n}$. After the measurement, he will obtain a guess y^n for the codeword chosen by Alice x^n and hence for the message m. The decoding map can thus be written as

$$\mathcal{D} : \mathfrak{D}(\mathcal{H}^{\otimes n}) \to \mathfrak{D}(\mathcal{H}_{\mathcal{M}}). \tag{11.20}$$

According to this scheme, the (average) error probability is

$$\begin{aligned} p_e &= \frac{1}{|\mathcal{M}|} \sum_m \Pr[\mathcal{D} \circ \mathcal{N}^{\otimes n} \circ \mathcal{E}(m) \neq m] \\ &= 1 - \frac{1}{|\mathcal{M}|} \sum_m \text{Tr}[\Xi_{x^n(m)}(\mathcal{N}^{\otimes n}(\rho_{x^n(m)}))], \end{aligned} \tag{11.21}$$

and the communication rate is $\frac{1}{n} \log |\mathcal{M}|$.

Definition 11.2.1. A communication rate $\mathtt{R}$ is said to be achievable if there exists a sequence (over n) of encoding/decoding maps such that

$$\lim_{n\to\infty}\frac{1}{n}\log|\mathcal{M}|=\mathtt{R},\quad\text{and}\quad\lim_{n\to\infty}p_e=0.$$

The sup among achievable communication rates defines the so-called product states classical capacity.

Then the following theorem, named HSW coding theorem after Holevo, Schumacher and Westmoreland, holds true.

Theorem 11.2.2 (HSW coding theorem). *The product states classical capacity of the memoryless quantum channel $\mathcal{N}$ is given by*

$$\begin{aligned}C_1(\mathcal{N})&=\max_{\{p_x,\rho_x\}}\chi(\{p_x,\mathcal{N}(\rho_x)\})\\&\equiv\max_{\{p_x,\rho_x\}}\left[S\left(\mathcal{N}\left(\sum_x p_x\rho_x\right)\right)-\sum_x p_x S(\mathcal{N}(\rho_x))\right].\end{aligned}$$

The subscript 1 on C refers to the fact that such a capacity is expressed in terms of a single channel use.

Proof. *Converse part.* Here, we prove that the rate $\mathtt{R}$ is upper bounded by $\max_{\{p_x,\rho_x\}}\chi(\{p_x,\mathcal{N}(\rho_x)\})$. Exploiting the fact that the channel is memoryless and the coding procedure uses product states, we can refer to (11.10) and use there the Holevo bound 9.2.1 (applied to the channel output) to get

$$\mathtt{R}\leq\frac{H_2(p_e)}{n(1-p_e)}+\frac{1}{1-p_e}\max_{\{p_x,\rho_x\}}\chi(\{p_x,\mathcal{N}(\rho_x)\}).$$

Taking the limit $p_e\to 0$ yields the desired inequality.

Direct part. Achievability of the rate $\chi(\{p_x,\mathcal{N}(\rho_x)\})$ is shown by resorting to random coding similar to Shannon's noisy channel coding theorem. We will require the following auxiliary result, stated without proof.

Lemma 11.2.3. *For any operators S and T in a Hilbert space $\mathcal{H}$, such that $0 \leq S \leq I$ and $T \geq 0$, the following relation holds true*

$$I - (S+T)^{-1/2} S (S+T)^{-1/2} \leq (1+c)(I-S) + (2+c+c^{-1})T, \tag{11.22}$$

with $c \in \mathbb{R}_+$.

Suppose that on each use of the quantum channel $\mathcal{N}$, a symbol $x \in \mathcal{X}$ is encoded into ρ_x with probability p_x and transmitted through the quantum channel:

$$\rho_x \mapsto \sigma_x = \mathcal{N}(\rho_x),$$

so that the average output reads $\bar{\sigma} = \sum_x p_x \sigma_x$.

Actually, Alice wants to send a message $m \in \mathcal{M}$ (with $|\mathcal{M}| = 2^{n\mathrm{R}}$) and encodes it into the codeword $\rho_{x_1} \otimes \rho_{x_2} \otimes \cdots \otimes \rho_{x_n}$ and the output of n channel uses will be

$$\sigma_m := \mathcal{N}^{\otimes n}(\rho_{x_1} \otimes \rho_{x_2} \otimes \cdots \otimes \rho_{x_n}). \tag{11.23}$$

Now, fix $\epsilon > 0$ and let P be the projector onto the ϵ-typical subspace of $\bar{\sigma}^{\otimes n}$. Consider also the average output entropy

$$\bar{S} := \sum_x p_x S(\sigma_x), \tag{11.24}$$

and define as P_m the projector onto the space spanned by eigenvectors of σ_m whose eigenvalues $\lambda_k(\sigma_m)$ satisfy the condition

$$\left| \frac{1}{n} \log \frac{1}{\lambda_k(\sigma_m)} - \bar{S} \right| \leq \epsilon, \quad \forall k. \tag{11.25}$$

Similarly to the theorem of typical sequences, for any $\delta > 0$ and sufficiently large n, we then have

$$\mathbb{E}[\mathrm{Tr}(P_m \sigma_m)] \equiv \sum_m p(x^n(m)) \mathrm{Tr}(P_m \sigma_m) \geq 1 - \delta, \tag{11.26}$$

where the expectation is taken with respect to the distribution over codewords (for a fixed message m) induced by random coding. Also, the dimension of the subspace onto which P_m projects can be at most $2^{n(\bar{S}+\epsilon)}$, hence

$$\mathbb{E}[\mathrm{Tr}(P_m)] \equiv \sum_m p(x^n(m))\mathrm{Tr}(P_m) \leq 2^{n(\bar{S}+\epsilon)}. \tag{11.27}$$

Bob tries to decode the message m by the POVM

$$E_m := \left(\sum_{m'} PP_{m'}P\right)^{-1/2} PP_mP \left(\sum_{m'} PP_{m'}P\right)^{-1/2}.$$

Loosely speaking, this POVM corresponds to checking to see if the channel output falls into the space on which P_m projects.

For a specific codebook $\{x^n(m')\}$ and the transmitted codeword $x^n(m)$, the probability of error is given by

$$\Pr(\text{error}|x^n(m), \{x^n(m')\}) = \mathrm{Tr}[(I - E_m)\sigma_m].$$

We now use Lemma 11.2.3 with $S = PP_mP$ and $T = \sum_{m' \neq m} PP_{m'}P$ to bound this by

$$\begin{aligned}\Pr(\text{error}|x^n(m), \{x^n(m')\}) &\leq (1+c)(1 - \mathrm{Tr}[PP_mP\sigma_m]) \\ &\quad + (2+c+c^{-1}) \sum_{m' \neq m} \mathrm{Tr}[PP_{m'}P\sigma_m].\end{aligned}$$

Averaging over all codebooks, but keeping the transmitted codeword m fixed, we find

$$\begin{aligned}\Pr(\text{error}|x^n(m)) &\leq (1+c)(1 - \mathrm{Tr}[PP_mP\sigma_m]) \\ &\quad + (2+c+c^{-1})(2^{nR} - 1) \\ &\quad \times \mathrm{Tr}\left[\sum_{m'} p(x^n(m'))PP_{m'}P\sigma_m\right].\end{aligned}$$

Averaging now in addition over the transmitted codeword $x^n(m)$, we obtain the upper bound

$$\Pr(\text{error}) \leq (1+c)\left(1 - \sum_m p(x^n(m))\mathrm{Tr}[PP_mP\sigma_m]\right)$$
$$+ (2+c+c^{-1})2^{nR}\mathrm{Tr}\left[\sum_{m'} p(x^n(m'))PP_{m'}P\right.$$
$$\left.\times \sum_m p(x^n(m))\sigma_m\right]. \tag{11.28}$$

Now, concerning the first term at right hand side of (11.28), we may note that

$$\begin{aligned}
&\sum_m p(x^n(m))\mathrm{Tr}[PP_mP\sigma_m] \\
&\quad = \sum_m p(x^n(m))\mathrm{Tr}[P\sigma_mP]\mathrm{Tr}\left[P_m\frac{P\sigma_mP}{\mathrm{Tr}(P\sigma_mP)}\right] \\
&\quad \geq \sum_m p(x^n(m))\mathrm{Tr}[P\sigma_mP](1-\delta) \\
&\quad = \mathrm{Tr}[P\bar{\sigma}^{\otimes n}P](1-\delta) \\
&\quad \geq (1-\delta)^2,
\end{aligned} \tag{11.29}$$

where we have exploited (11.26) and the fact that the density operator $\frac{P\sigma_mP}{\mathrm{Tr}(P\sigma_mP)}$ has support within that of σ_m.

Concerning the second term at right hand side of (11.28), we may note that

$$\begin{aligned}
&\mathrm{Tr}\left[\sum_{m'} p(x^n(m'))PP_{m'}P\sum_m p(x^n(m))\sigma_m\right] \\
&\quad = \mathrm{Tr}\left[P\bar{\sigma}^{\otimes n}P\sum_{m'} p(x^n(m'))P_{m'}\right].
\end{aligned}$$

However,

$$P\bar{\sigma}^{\otimes n}P \leq 2^{-n(S(\bar{\sigma})-\epsilon)}I$$

and using (11.27), we arrive at

$$\operatorname{Tr}\left[\sum_{m'} p(x^n(m'))PP_{m'}P\sum_{m} p(x^n(m))\sigma_m\right] \leq 2^{nR-n(S(\bar{\sigma})-\bar{S}-2\epsilon)}. \tag{11.30}$$

Inserting expressions (11.29) and (11.30) into (11.28), we finally find

$$\Pr(\text{error}) \leq 2(1+c)\delta + (2+c+c^{-1})2^{nR-n(S(\bar{\sigma})-\bar{S}-2\epsilon)}, \tag{11.31}$$

concluding the proof. □

Alice can indeed use a more general encoding strategy. Instead of using product states, she can use states which are *entangled* across channel uses. Then she will associate to each message $m \in \mathcal{M}$ a state $\rho_{x^n(m)} \in \mathfrak{D}(A^n)$.

Definition 11.2.4. The maximum achievable classical communication rate obtained by allowing this more general encoding strategy is called classical capacity of the quantum channel $\mathcal{N}$, denoted $C(\mathcal{N})$.

Theorem 11.2.5. *The classical capacity of the memoryless quantum channel $\mathcal{N}$ is given by*

$$C(\mathcal{N}) = \lim_{n\to\infty} \frac{1}{n} \max_{\{p_x, \rho_x^{(n)}\}} \chi(\{p_x, \mathcal{N}^{\otimes n}(\rho_x^{(n)})\}).$$

This coding theorem is a straightforward generalization of the HSW theorem. Indeed, if we look at n uses of the qubit channel, $\mathcal{N}^{\otimes n}$, as a single use of channel acting on n qubits, and then optimize over n, we obtain the coding theorem for the classical capacity of the quantum channel involving the regularized limit over channel uses, that is $C = \lim_{n\to\infty} \frac{1}{n} C_1(\mathcal{N}^{\otimes n})$.

It can be easily proven that, in computing the maximum over ensembles, it is always possible to consider only ensembles of pure states $\{p_x, |\psi_x^{(n)}\rangle\langle\psi_x^{(n)}|\}$. In fact, for the first term in the χ quantity we have

$$S\left(\mathcal{N}\left(\sum_{x,i} p_x \lambda_i^x |\varphi_i^x\rangle\langle\varphi_i^x|\right)\right) = S\left(\mathcal{N}\left(\sum_k q_k |\varphi_k\rangle\langle\varphi_k|\right)\right),$$

with $k = \{x, i\}$ and q_k a probability distribution. Furthermore, for the second term in the χ quantity we have

$$\begin{aligned}
\sum_x p_x S(\mathcal{N}(\rho_x)) &= \sum_x p_x S\left(\mathcal{N}\left(\sum_i \lambda_i^x |\varphi_i^x\rangle\langle\varphi_i^x|\right)\right) \\
&= \sum_x p_x S\left(\sum_i \lambda_i^x \mathcal{N}(|\varphi_i^x\rangle\langle\varphi_i^x|)\right) \\
&\geq \sum_{x,i} p_x \lambda_i^x S(\mathcal{N}(|\varphi_i^x\rangle\langle\varphi_i^x|)) \\
&= \sum_k q_k S(\mathcal{N}(|\varphi_k\rangle\langle\varphi_k|)),
\end{aligned}$$

with $k = \{x, i\}$ and q_k a probability distribution.

The classical capacity of a quantum channel is in general *superadditive*, that is

$$C(\mathcal{N}) > C_1(\mathcal{N}), \tag{11.32}$$

although, for several classes of channels it has been proven that $C = C_1$. In particular, the equality holds for entanglement breaking channels, for all depolarizing channels and for unital qubit channels.

Example. The capacity C_1 of the qubit depolarizing channel

$$\mathcal{N}(\rho) = q\rho + (1-q)\frac{I}{2}.$$

We have to compute

$$C_1 = \max_{\{p_i, |\psi_i\rangle\}} S\left(\sum_i p_i \mathcal{N}(|\psi_i\rangle\langle\psi_i|)\right) - \sum_i p_i S(\mathcal{N}(|\psi_i\rangle\langle\psi_i|)).$$

For a pure input state formed with the state vector $|\psi_i\rangle = \alpha_i|0\rangle + \beta_i|1\rangle$, the density matrix of the output state is

$$\mathcal{N}(|\psi_i\rangle\langle\psi_i|) = q|\psi_i\rangle\langle\psi_i| - (1-q)\frac{I}{2}$$
$$= \begin{pmatrix} q|\alpha_i|^2 + \dfrac{1-q}{2} & q\alpha_i\beta_i^* \\ q\alpha_i\beta_i^* & q|\beta_i|^2 + \dfrac{1-q}{2} \end{pmatrix},$$

whose eigenvalues are $(1 \pm q)/2$, hence

$$S[\mathcal{N}(|\psi_i\rangle\langle\psi_i|)] = H_2\left(\frac{1-q}{2}\right),$$

independently of α_i and β_i. Now note that by linearity,

$$S\left(\sum_i p_i \mathcal{N}(|\psi_i\rangle\langle\psi_i|)\right) = S\left(\mathcal{N}\left(\sum_i p_i|\psi_i\rangle\langle\psi_i|\right)\right),$$

and using the fact that the channel is unital, the maximum is achieved when $\sum_i p_i|\psi_i\rangle\langle\psi_i| = I/2$. This can be realized in several ways, e.g., by taking $|\psi_0\rangle = |0\rangle$, $|\psi_1\rangle = |1\rangle$ with $p_0 = p_1 = 1/2$, yielding

$$C_1 = 1 - H_2\left(\frac{1-q}{2}\right). \tag{11.33}$$

For the qubit depolarizing channel, it can be proved that $C = C_1$.

11.3 Entanglement-assisted classical capacity

Consider now Alice sending classical information encoded into quantum states (e.g. qubits) through a quantum *memoryless* channel $\mathcal{N} : \mathfrak{D}(\mathcal{H}_A) \rightarrow \mathfrak{D}(\mathcal{H}_B)$, with $\mathcal{H}_A \simeq \mathcal{H}_B \simeq \mathbb{C}^2$, but this time exploiting pre-shared entanglement with Bob, that is considered the generalization of the dense coding protocol discussed in Section 9.4.2 to the case of non-ideal quantum channel $\mathcal{N}$ between Alice and Bob.

The code in this case is specified by considering a set $\mathcal{M}$ of possible messages, a family of encoding maps

$$\{\mathcal{E}_m : \mathfrak{D}(\mathcal{H}_U) \to \mathfrak{D}(\mathcal{H}_A^{\otimes n})\}_m, \tag{11.34}$$

acting on a state ω_{UV} shared between Alice and Bob before the communication, and resulting in the signal states

$$(\rho_m)_{A^n V} = (\mathcal{E}_m \otimes \mathrm{id})\omega_{UV}.$$

After the n channel uses, Bob thus receives the state

$$(\mathcal{N}^{\otimes n} \otimes \mathrm{id})\rho_m \in \mathfrak{D}(B^n V).$$

On it, Bob will perform a decoding measurement with POVM elements $\{\Xi_m\}_m$, corresponding to his estimate m of the sent message. According to this scheme, the (average) error probability is

$$\begin{aligned} p_e &= \frac{1}{|\mathcal{M}|}\sum_m \Pr[\mathcal{D} \circ (\mathcal{N}^{\otimes n} \otimes \mathrm{id}) \circ (\mathcal{E}_m \otimes \mathrm{id})\omega \neq m] \\ &= 1 - \frac{1}{|\mathcal{M}|}\sum_m \mathrm{Tr}[\Xi_m((\mathcal{N}^{\otimes n} \circ \mathcal{E}_m \otimes \mathrm{id})\omega)], \end{aligned} \tag{11.35}$$

and the communication rate is $\frac{1}{n}\log|\mathcal{M}|$.

Definition 11.3.1. A communication rate $\mathtt{R}$ is said to be achievable if there exists a sequence (over n) of encoding maps families $\{\mathcal{E}_m\}_m$ and decoding maps $\mathcal{D}$ such that

$$\lim_{n\to\infty}\frac{1}{n}\log|\mathcal{M}| = \mathtt{R}, \quad \text{and} \quad \lim_{n\to\infty} p_e = 0.$$

The sup among achievable communication rates defines the so-called entanglement-assisted classical capacity.

Theorem 11.3.2 (Coding theorem for entanglement-assisted classical capacity). *The entanglement assisted classical capacity of a memoryless quantum channel* $\mathcal{N} : \mathfrak{D}(\mathbb{C}^2) \to \mathfrak{D}(\mathbb{C}^2)$ *is given by*

$$C_E(\mathcal{N}) = \max_\rho[S(\rho) + S(\mathcal{N}(\rho)) - S((\mathcal{N} \otimes \mathrm{id})|\psi_\rho\rangle\langle\psi_\rho|)],$$

where $|\psi_\rho\rangle\langle\psi_\rho|$ *is a purification of* ρ.

Proof. *Converse part.* From (11.5), using Theorem **??**, we get

$$\mathtt{R} \leq \frac{H_2(p_e)}{n(1-p_e)} + \frac{I(X^n:Y^n)}{n(1-p_e)} \leq \frac{H_2(p_e)}{n(1-p_e)} + \frac{I(A^n;R^n)}{n(1-p_e)},$$

where $I(A^n;R^n)$ is the quantum mutual information between input A^n and reference R^n. Since the latter quantity is subadditive (Theorem 6.4.3), we further have

$$\mathtt{R} \leq \frac{H_2(p_e)}{n(1-p_e)} + \frac{\sum_{i=1}^n I(A_i;R_i)}{n(1-p_e)} \leq \frac{H_2(p_e)}{n(1-p_e)} + \frac{\max_\rho I(A;R)}{(1-p_e)}.$$

Taking the limit $p_e \to 0$ yields the desired inequality.

Direct part. It remains to prove the achievability, or equivalently that

$$\mathtt{R} \geq [S(\rho) + S(\mathcal{N}(\rho)) - S((\mathcal{N}\otimes \mathrm{id})|\psi_\rho\rangle\langle\psi_\rho|)], \tag{11.36}$$

holds also true.

It is easy to show equation (11.36) for the special case of $\rho = I/2$ and $|\psi_\rho\rangle$ maximally entangled state vector. To this end, we employ the Pauli matrices $U_{0,0} = I, U_{0,1} = \sigma^Z, U_{1,0} = \sigma^X, U_{1,1} = \sigma^X\sigma^Z$. Then, to achieve the rate at right hand side of equation (11.36) Alice and Bob start sharing a (four-dimensional) maximally entangled state vector $|\psi_\rho\rangle$, then Alice applies one of the 4 transformations $U_{j,k}$ on her part of $|\psi_\rho\rangle$ and sends it through the channel $\mathcal{N}$. Bob gets one of the four states $(\mathcal{N}\otimes \mathrm{id})(\mathcal{U}_{j,k}\otimes \mathrm{id})|\psi_\rho\rangle\langle\psi_\rho|$. Here, $\mathcal{U}_{j,k}(\rho) = U_{i,j}\rho U_{i,j}^\dagger$. The average overall $\{U_{j,k}\}$ yields

$$\sum_{j,k=0}^{1} (\mathcal{N}\otimes \mathrm{id})(\mathcal{U}_{j,k}\otimes \mathrm{id})|\psi_\rho\rangle\langle\psi_\rho|$$
$$= \mathcal{N}(\mathrm{Tr}_B|\psi_\rho\rangle\langle\psi_\rho|) \otimes \mathrm{Tr}_A|\psi_\rho\rangle\langle\psi_\rho| = \mathcal{N}(\rho)\otimes\rho,$$

where $\rho = I/2$. The entropy of this quantity gives the first two terms of (11.36). The entropy of each of the four states $(\mathcal{N}\otimes \mathrm{id})(\mathcal{U}_{j,k}\otimes \mathrm{id})|\psi_\rho\rangle\langle\psi_\rho|$ is $S(\mathcal{N}\otimes \mathrm{id})(|\psi_\rho\rangle\langle\psi_\rho|)$ since each of the $(\mathcal{U}_{j,k}\otimes \mathrm{id})|\psi_\rho\rangle\langle\psi_\rho|$ is a purification of ρ. This entropy gives the third term at right hand side of equation (11.36).

The next step is to show that equation (11.36) holds for an arbitrary ρ. To this end, we need the following.

Lemma 11.3.3. *Let $\mathcal{N}$ be a quantum channel and ρ a density operator on its input space, then we can find a sequence (over n) of typical subspaces $\mathcal{T}_\rho(n, \epsilon)$ corresponding to ρ such that, if $\pi_{\mathcal{T}_\rho}$ is the density operator proportional to the projection onto $\mathcal{T}_\rho$, it holds*

$$\lim_{n\to\infty} \frac{1}{n} S(\mathcal{N}^{\otimes n}(\pi_{\mathcal{T}_\rho})) = S(\mathcal{N}(\rho)).$$

This lemma extends the well-known fact about typical subspace that $\lim_{n\to\infty} \frac{1}{n} S(\pi_{\mathcal{T}_\rho}) = S(\rho)$.

Proof of the Lemma (sketch). For the $\leq$ part, one can resort to Schumacher's compression Theorem 8.2.2 to show that a source producing states with average density operator $\mathcal{N}^{\otimes n}(\pi_{\mathcal{T}_\rho})$ can be faithfully compressed into $nS(\mathcal{N}(\rho)) + o(n)$ qubit.

For the $\geq$ part, one has to devise a communication protocol which transmits a classical message containing $nS(\mathcal{N}(\rho)) - o(n)$ bits using pure states and then apply the HSW theorem to this protocol to deduce a lower bound to the entropy $S(\mathcal{N}^{\otimes n}(\pi_{\mathcal{T}_\rho}))$. □

Coming back to the proof of the theorem, consider the channel $\widetilde{\mathcal{N}}$ complementary to $\mathcal{N}$ (see Definition 6.3.4). Applying Lemma 11.3.3 to this CPTP map gives

$$\lim_{n\to\infty} \frac{1}{n} S(\widetilde{\mathcal{N}}^{\otimes n}(\pi_{\mathcal{T}_\rho})) = S(\widetilde{\mathcal{N}}(\rho)).$$

Thus if we consider the quantity

$$\frac{1}{n}[S(\pi_{\mathcal{T}_\rho}) + S(\mathcal{N}^{\otimes n}(\pi_{\mathcal{T}_\rho})) - S(\widetilde{\mathcal{N}}^{\otimes n}(\pi_{\mathcal{T}_\rho}))],$$

we see that it converges to

$$S(\rho) + S(\mathcal{N}(\rho)) - S\left(\widetilde{\mathcal{N}}(\rho)\right).$$

But for the channel $\widetilde{\mathcal{N}}$, we have

$$S(\widetilde{\mathcal{N}}(\rho)) = S(\mathcal{N} \otimes \mathrm{id})(|\psi_\rho\rangle\langle\psi_\rho|),$$

hence the desired final result. □

Remark. Due to the subadditivity of the quantum mutual information (Theorem 6.4.3), the expression for C_E in Theorem 11.3.2

does not require the regularization over the channel uses, i.e. the quantity C_E is additive.

Remark. While for a qubit channel, $0 \leq C(\mathcal{N}) \leq 1$, it is easy to see that $0 \leq C_E(\mathcal{N}) \leq 2$. Obviously, also $C(\mathcal{N}) \leq C_E(\mathcal{N})$.

11.4 Quantum capacity

We have seen that quantum error correction allows to undo the irreversible changes caused by noisy quantum channels. Hence, quantum information can be transmitted through noisy quantum channels.

Let us consider a quantum *memoryless* channel $\mathcal{N} : \mathfrak{D}(\mathcal{H}_A) \to \mathfrak{D}(\mathcal{H}_B)$. Define a quantum code by encoding and decoding maps

$$\mathcal{E} : \mathfrak{D}(\mathcal{H}) \to \mathfrak{D}(\mathcal{H}_A^{\otimes n}), \tag{11.37}$$

$$\mathcal{D} : \mathfrak{D}(\mathcal{H}_B^{\otimes n}) \to \mathfrak{D}(\mathcal{H}), \tag{11.38}$$

where $\mathcal{H}$ is a Hilbert space such that $|\mathcal{H}| \leq 2^n$. The goal is to transmit quantum states as faithfully as possible on average, which can be expressed compactly as the demand that a maximally entangled state $|\Phi\rangle \in \mathcal{H} \otimes \mathcal{H}'$ between $\mathcal{H}$ and a reference system $\mathcal{H}'$ is recovered with high fidelity. The (*average*) *error* (sometimes called *indelity*) is given by

$$p_e = 1 - F\left(\Phi, \left(\mathcal{D} \circ \mathcal{N}^{\otimes n} \circ \mathcal{E} \otimes \mathrm{id}_{\mathcal{H}'}\right) \Phi\right), \tag{11.39}$$

and the rate results $\frac{1}{n} \log |\mathcal{H}|$.

Definition 11.4.1. A communication rate $\mathtt{R}$ is said to be achievable if there exists a sequence (over n) of encoding/decoding maps such that

$$\lim_{n\to\infty} \frac{1}{n} \log |\mathcal{H}| = \mathtt{R}, \quad \text{and} \quad \lim_{n\to\infty} p_e = 0.$$

The supremum among achievable communication rates defines the quantum capacity of the channel, denoted by $Q(\mathcal{N})$.

Theorem 11.4.2 (Coding theorem for quantum capacity). *The quantum capacity of a memoryless qubit channel $\mathcal{N}$ is given by*

$$Q(\mathcal{N}) = \lim_{n\to\infty} \frac{1}{n} \max_{\rho^{(n)}} I_{\mathrm{coh}}(\rho^{(n)}, \mathcal{N}^{\otimes n}),$$

where I_{coh} is the coherent information.

Proof. *Converse part.* The upper bound for $\mathtt{R}$ follows from Theorem 9.3.1 when applied to n channel uses, i.e.

$$\begin{aligned}\frac{S(\rho^{(n)})}{n} &\leq \frac{I_{\mathrm{coh}}(\rho^{(n)}, \mathcal{N}^{\otimes n})}{n} + 2\frac{H_2(p_e)}{n} + 2\frac{p_e}{n}\log(|\mathcal{H}|-1)\\ &\leq \frac{\max_{\rho^{(n)}} I_{\mathrm{coh}}(\rho^{(n)}, \mathcal{N}^{\otimes n})}{n} + 2\frac{H_2(p_e)}{n} + 2\frac{p_e}{n}\log(|\mathcal{H}|-1).\end{aligned}$$

Specifying the above for the maximally mixed input state gives

$$\mathtt{R} \leq \frac{\max_{\rho^{(n)}} I_{\mathrm{coh}}(\rho^{(n)}, \mathcal{N}^{\otimes n})}{n} + 2\frac{H_2(p_e)}{n} + 2\frac{p_e}{n}\log(|\mathcal{H}|-1),$$

which yields the desired inequality as soon as $p_e \to 0$.

Direct part (sketch). To prove that $\frac{1}{n}\max_{\rho^{(n)}} I_{\mathrm{coh}}(\rho^{(n)}, \mathcal{N}^{\otimes n})$ is also an achievable rate, first note that besides simulating a noiseless channel on arbitrary inputs, an equivalent task for conveying quantum information through a channel with arbitrarily small error is to establish maximal entanglement between sender and receiver.

An entanglement generation code of rate $\mathtt{R}$ for $\mathcal{N}$ consists of two isomorphic $2^{n\mathtt{R}}$-dimensional Hilbert spaces $\mathcal{H}_{R'}$ and $\mathcal{H}_R$ with associated maximally entangled state vector $|\Phi\rangle_{R'R}$, an encoding state $|\Upsilon\rangle_{R'A^n}$ and a decoding map $\mathcal{D} : \mathfrak{D}(\mathcal{H}_B^{\otimes n}) \to \mathfrak{D}(\mathcal{H}_R)$ so that Eq. (11.39) is replaced by

$$p_e = 1 - F(|\Phi\rangle_{R'R}, (\mathrm{id}_{R'} \otimes (\mathcal{D} \circ \mathcal{N}^{\otimes n}))(\Upsilon_{R'A^n})). \qquad (11.40)$$

The fact that every $0 \leq \mathtt{R} \leq I_{\mathrm{coh}}(\rho, \mathcal{N})$ is an achievable rate can be proved by means of the following lemma that bounds how well entanglement between input A and reference R' systems can be recovered by how decoupled R' is from the environment E of the channel.

Lemma 11.4.3 (Decoupling from environment). *Let* $V_{\mathcal{N}}^{A\to BE} : \mathcal{H}_A \to \mathcal{H}_B \otimes \mathcal{H}_E$ *be a (Stinespring) dilation of the channel* $\mathcal{N} : \mathfrak{D}(\mathcal{H}_A) \to \mathfrak{D}(\mathcal{H}_B)$. *Fixing a Hilbert space* $\mathcal{H}_{R'}$ *satisfying* $|\mathcal{H}_{R'}| \leq |\mathcal{H}_A^{\otimes n}|$, *let* $|\Phi\rangle_{R'A^n}$ *be maximally entangled and set*

$$|\psi\rangle_{R'B^nE^n} := (V_{\mathcal{N}}^{A\to BE})^{\otimes n}|\Phi\rangle_{R'A^n}.$$

Then, there is a decoding map $\mathcal{D} : \mathfrak{D}(\mathcal{H}_B^{\otimes n}) \to \mathfrak{D}(\mathcal{H}_R)$ *satisfying*

$$\begin{aligned} &F(|\Phi\rangle_{R'R}, (\mathrm{id}_{R'} \otimes (\mathcal{D} \circ \mathcal{N}^{\otimes n})(\Upsilon_{R'A}))) \\ &\quad \geq 1 - \|\psi_{R'E^n} - \pi_{R'} \otimes \psi_{E^n}\|_1, \end{aligned} \tag{11.41}$$

where π *denotes the maximally mixed state and* ψ *the density operator coming from* $|\psi\rangle\langle\psi|$ *by partial trace.*

Proof of Lemma. By Uhlmann's Theorem 5.3.2, there is a purification $|\psi'\rangle_{R'B^nE^n}$ of $\pi_R \otimes \psi_{E^n}$ satisfying

$$|\langle\psi|\psi'\rangle|^2 = F(\psi_{R'E^n}, \pi_{R'} \otimes \psi_{E^n}) \geq 1 - \epsilon.$$

Because ${\psi'}_{R'} = \pi_{R'}$ is maximally mixed, it is purified by a maximally entangled state vector $|\Phi\rangle_{R'R}$. Furthermore, since $\pi_{R'} \otimes \psi_{E^n}$ is a product state, it must have a purification which is a tensor product of pure states. Therefore, there is another Hilbert space $\mathcal{H}_K$, a state vector $|\xi\rangle_{KE^n}$ and an isometry $W^{B^n\to RK} : \mathcal{H}_B^{\otimes n} \to \mathcal{H}_R \otimes \mathcal{H}_K$ under which

$$W^{B^n\to RK}|\psi'\rangle_{R'B^nE^n} = |\Phi\rangle_{R'R}|\xi\rangle_{KE^n}.$$

Combining monotonicity of fidelity, $F(\phi, \sigma) \leq F(\mathcal{N}(\phi), \mathcal{N}(\sigma))$, with the relation $F(\phi, \sigma) \geq 1 - \|\phi - \sigma\|_1$, this implies that the decoding $\mathcal{D}^{B^n\to R} := \mathrm{Tr}_K W^{B^n\to RK}$ satisfies (11.41) as required. □

Coming back to the proof of Theorem 11.4.2, the idea is to select coding subspaces such that if they are sufficiently small, any data contained within them will with high probability be decoupled from the noisy channel's environment.

Given the channel $\mathcal{N}^{A\to B}$ and a density matrix φ_A, fix a (Stinespring) dilation $V_{\mathcal{N}}^{A\to BE}$ and a purification $|\varphi\rangle_{A'A}$, where $\mathcal{H}_{A'} \simeq \mathcal{H}_A$. Feeding part of the purification through the Stinespring dilation gives the state vector $|\varphi\rangle_{A'BE} = (I_{A'} \otimes V_{\mathcal{N}})|\varphi\rangle_{A'A}$. Fixing $\delta, \epsilon > 0$, we take a subspace $\mathcal{H}_{R'} \subset \mathcal{T}_{\varphi A'}$ of the typical subspace $\mathcal{T}_{\varphi A'}$ corresponding to $\varphi_{A'}$ and we let $\mathcal{H}_R \subset \mathcal{H}_A^{\otimes n}$ be isomorphic to $\mathcal{H}_{R'}$.

For a maximally entangled state vector $|\Phi\rangle_{R'R}$ and for each unitary $U : \mathcal{T}_{\varphi A'} \to \mathcal{T}_{\varphi A'}$, define

$$\begin{aligned}
|\psi_U\rangle_{R'B^nE^n} &:= (U \otimes I_{B^n} \otimes I_{E^n})(I_{R'} \otimes V_{\mathcal{N}}^{\otimes n})|\Phi\rangle_{R'R} \\
&= (U \otimes I_{B^n} \otimes I_{E^n})(I_{R'} \otimes V_{\mathcal{N}}^{\otimes n} U^\dagger U)|\Phi\rangle_{R'R} \\
&= (I_{R'} \otimes V_{\mathcal{N}}^{\otimes n} U^\dagger)|\tilde{\Phi}\rangle_{R'R},
\end{aligned}$$

where $|\tilde{\Phi}\rangle_{R'R} := (U \otimes U)|\Phi\rangle_{R'R}$ is still a maximally entangled state vector.

Each of the state vectors $(I_{R'} \otimes U^\dagger)|\tilde{\Phi}\rangle_{R'R}$ can be identified with a possible choice for the encoding $|\Upsilon\rangle_{R'A^n}$, where R' is maximally entangled with a particular subspace of A^n.

Then it is possible to show (not reported here) that

$$\int_{\mathrm{U}(R')} \|(\psi_U)_{R'E^n} - \pi_{R'} \otimes (\psi_U)_{E^n}\|_1 \, dU \leq 3\epsilon,$$

provided the transmission rate satisfies $\mathtt{R} < I_{\text{coh}}(\varphi_A, \mathcal{N}) - 4\delta$.

Given that U represents the encoding process, this is analogous to averaging over random codes, where now the average is performed over the group $\mathrm{U}(R')$ of unitaries on R' and with the Haar measure of that group (see appendix).

The above result implies the existence of a U such that R' and E^n are sufficiently decoupled in the corresponding state, namely

$$\|(\psi_U)_{R'E^n} - \pi_{R'} \otimes (\psi_U)_{E^n}\|_1 \leq 3\epsilon.$$

Using the state vector $|\Upsilon\rangle_{R'A^n} = (I \otimes U^\dagger)|\tilde{\Phi}\rangle_{R'R}$ for encoding, we may now invoke Lemma 11.4.3 to infer the existence of a good decoder, namely a code for entanglement transmission through $\mathcal{N}$ with fidelity greater than or equal to $1 - 3\epsilon$.

Since δ can be taken to be arbitrarily small, there is for every rate satisfying

$$\mathtt{R} < \min\left\{\frac{1}{n}\log|\mathcal{T}_{\varphi A'}|, I_{\text{coh}}(\varphi_A, \mathcal{N})\right\},$$

a sequence of entanglement generation codes for $\mathcal{N}^{A\to B}$ with error probability going to zero. Moreover, from typicality arguments it holds

$$|\mathcal{T}_{\varphi A'}| \geq 2^{nS(\varphi_{A'})-\iota_n(\delta)} \tag{11.42}$$

where $\iota_n(\delta) \to 0$ as $n \to \infty$, so that $\frac{1}{n}\log|\mathcal{T}_{\varphi_{A'}}|$ can be arbitrarily close to $S(\varphi_{A'}) = S(\varphi_A)$ (since $|\varphi_{A'A}\rangle\langle\varphi_{A'A}|$ is pure) and thus greater than $I_{\text{coh}}(\varphi_A, \mathcal{N})$ (according to Eq. (6.16)). □

Remark. We can define

$$Q_1(\mathcal{N}) := \max\{0, \max_\rho I_{\text{coh}}(\rho, \mathcal{N})\}, \tag{11.43}$$

so that it results

$$Q = \lim_{n\to\infty}\frac{1}{n}Q_1(\mathcal{N}^{\otimes n}).$$

Clearly, Q_1 is a lower bound on the quantum capacity, which quite generally results *superadditive*, i.e. $Q \geq Q_1$.

Remark. It is obvious that $0 \leq Q \leq 1$. The quantum capacity is always smaller or equal to the classical one, i.e. $Q(\mathcal{N}) \leq C(\mathcal{N})$.

Example. The quantum capacity of erasure channel. Letting p be the erasure probability, it holds trivially that $C = 1-p$, and one might be tempted to think that Q is the same. However, Alice sends n qubits, Bob receives $\approx (1-p)n$ of them while $\approx pn$ will be intercepted by someone else, say Charlie. For $p > 1/2$, Charlie ends up with more qubits than Bob. Then, if $Q > 0$ for $p \geq 1/2$, it would mean that Bob could recover the quantum state sent by Alice as well as Charlie could do. This is in contradiction to no-cloning Theorem. Hence, it must be that $Q = 0$ for $p \geq 1/2$. Actually, it is known that $Q = 1-2p$. Finally, note that for the erasure channel, the entanglement-assisted capacity results $C_E = 2(1-p)$, so as to have $Q \leq C \leq C_E$.

Remark. Similarly, one can use unbounded pre-shared entanglement to assist the transmission of quantum information. In this case, one is led to the definition of *entanglement-assisted quantum capacity.* It can be proven that $Q_E = \frac{1}{2}C_E$.

The possibility of quantum capacity becoming additive, i.e. $Q(\mathcal{N}) = Q_1(\mathcal{N})$, is related to an interesting property of the channel $\mathcal{N}$, that is its degradability.

Definition 11.4.4. A quantum channel $\mathcal{N}$ is *degradable* if one can recover the final environment state $\widetilde{\mathcal{N}}(\rho)$ just by applying a third CPTP map to the output system state. More formally, a degradable channel $\mathcal{N}$ is such that there exists a CPTP map $\mathcal{K}$ satisfying

$$\widetilde{\mathcal{N}} = \mathcal{K} \circ \mathcal{N}.$$

Similarly, a channel is called *anti-degradable* when the opposite relation holds

$$\mathcal{N} = \mathcal{K} \circ \widetilde{\mathcal{N}},$$

for some CPTP map $\mathcal{K}$.

It turns out that qubit channels having two Kraus operators of the form (in the canonical basis)

$$K_0 = \begin{pmatrix} \cos\theta & 0 \\ 0 & \cos\phi \end{pmatrix}, \quad K_1 = \begin{pmatrix} 0 & \sin\phi \\ \sin\theta & 0 \end{pmatrix}, \tag{11.44}$$

are degradable for $\frac{1-\sin^2\theta}{1-\sin^2\phi} \geq 0$ and antidegradable otherwise.

In the Stinespring representation, these maps describe situations in which the qubit system interacts with only one qubit environment.

Exercises

1. Consider the classical channel described by the following transition matrix:

$$P_{Y|X} = \begin{pmatrix} 1 & 0 \\ 1/2 & 1/2 \end{pmatrix},$$

which is known as (binary) Z channel. Discuss analogies with qubit amplitude damping channel. Find the capacity of the Z channel and the maximizing input probability mass function.

2. Let the ensemble that achieves $C_1(\mathcal{N})$ be $\{\tilde{p}_x, \tilde{\rho}_x\}_x$. Then show that

$$C_1(\mathcal{N}) = \sum_x \tilde{p}_x S(\mathcal{N}(\tilde{\rho}_x) \| \mathcal{N}(\tilde{\rho})),$$

where $\tilde{\rho} = \sum_x \tilde{p}_x \tilde{\rho}_x$.

3. Find the classical capacity of a generic Pauli channel $\mathcal{N}(\rho) = p_0\rho + p_1 \sigma^X \rho\sigma^X + p_2 \sigma^Y \rho\sigma^Y + p_3 \sigma^Z \rho\sigma^Z$. Also, compute the entanglement-assisted capacity in case it reduces to a depolarizing channel.

4. Compute the quantum capacity of a qubit amplitude damping channel acting as

$$\begin{aligned}\mathcal{N}(\rho) &= (|0\rangle\langle 0| + \sqrt{\eta}|1\rangle\langle 1|)\rho(|0\rangle\langle 0| + \sqrt{\eta}|1\rangle\langle 1|)^\dagger \\ &\quad + (\sqrt{1-\eta}|0\rangle\langle 1|)\rho(\sqrt{1-\eta}|0\rangle\langle 1|)^\dagger\end{aligned}$$

with $\eta \in [0, 1]$.

5. Find channels $\mathcal{N}_1$ and $\mathcal{N}_2$ with zero quantum capacity, i.e. $Q(\mathcal{N}_1) = Q(\mathcal{N}_2) = 0$, for which, by parallel use of the two communication lines $\mathcal{N}_1 \otimes \mathcal{N}_2$, it becomes possible to transmit quantum information, i.e. $Q(\mathcal{N}_1 \otimes \mathcal{N}_2) > 0$.

Chapter 12

ENTANGLEMENT MANIPULATION

As entanglement is a resource recognized to be useful for quantum communication (see, e.g. Quantum teleportation or Superdense coding), it is necessary to develop in this chapter a theory aimed at characterizing such a resource, that is, and how it can be transformed, and how it can be quantified.

12.1 Bipartite entanglement transformations

The analysis of entanglement as a resource leads to the problem of understanding which set of transformations can be implemented by Alice and Bob using only *Local Operations and Classical Communication* (LOCC) as physical elements. The set of LOCC transformations includes local measurements, local CPTP maps, local state preparation, unlimited classical communication from Alice to Bob and from Bob to Alice, and any concatenation of these operations.

12.1.1 *Local operations and classical communications*

We consider a pair of pure states with state vectors $|\psi\rangle_{AB}$, $|\phi\rangle_{AB}$, and ask under which conditions Alice and Bob can deterministically transform $|\psi\rangle_{AB}$ into $|\phi\rangle_{AB}$ by only acting on them with LOCC.

Theorem 12.1.1. *Any LOCC transformation of $|\psi\rangle_{AB}$ to $|\phi\rangle_{AB}$ can be realized by letting Alice make a the local measurements, send (one-way) the result to Bob, who applies a conditional unitary.*

Proof. Let us recall that any pure state can be written according to the Schmidt decomposition:

$$|\psi\rangle_{AB} = \sum_k \sqrt{\lambda_k}|k\rangle_A|k\rangle_B.$$

First, note that local unitaries cannot change the Schmidt coefficients λ_k, and that on the other hand the Schmidt bases can be changed to arbitrary local orthogonal bases. Then, suppose that Bob makes a measurement associated to the POVM elements $\{E^J = M^{J\dagger} M^J\}_J$. By expressing the measurement operators M^J in the Schmidt basis (on Bob's side) we have $M^J = M^J_{hk}|h\rangle_B\langle k|$, then the post-measurement state vector is

$$|\psi^J\rangle_{AB} = \sum_{hk} M^J_{hk}\sqrt{\lambda_k}|k\rangle_A|h\rangle_B,$$

with probability $p_J = \sum_{hk} \lambda_k |M^J_{hk}|^2$.

But, on the other hand, Alice could make a measurement with operator $M^J_{hk}|h\rangle_A\langle k|$, transforming with equal probability $p_J = \sum_{hk} \lambda_k |M^J_{hk}|^2$ the state vector to

$$|\psi'^J\rangle_{AB} = \sum_{hk} M^J_{hk}\sqrt{\lambda_k}|h\rangle_A|k\rangle_B,$$

which differs only by exchanging $|k\rangle_A \leftrightarrow |k\rangle_B$. This state vector has the same Schmidt coefficients, hence can be obtained from the previous one by applying local unitaries. That proves that all the measurements can be performed by Alice. What remains is that Alice sends the measurement result to Bob (classical communication), then Bob will operate a local unitary (or CPTP) transformation on his side. □

Accordingly, any LOCC protocol can be realized by letting Alice make all the measurements, and Bob the conditional unitaries. Since both kind of operations are performed locally, the product of POVM operators on Alice's side being a POVM operator on Alice's side, and the product of unitaries on Bob's side still being a unitary on Bob's side, it follows that a sequence of n rounds (Alice's measurements, communication to Bob, unitary by Bob) can be rewritten as a single round.

Majorization, i.e. a relation of partial ordering between probability distribution, allows to characterize pure states that can be transformed by LOCC.

Definition 12.1.2. Given two n-component real vectors $x = (x_1, x_2, \ldots, x_n)$, $y = (y_1, y_2, \ldots, y_n)$ arranged in decreasing order, i.e. $x_1 \geq x_2 \geq \cdots x_n$ and $y_1 \geq y_2 \geq \cdots y_n$, we say that x is majorized by y (and denote it by $x \prec y$) if

$$\sum_{k=1}^{d} x_d \leq \sum_{k=1}^{d} y_d,$$

for all $d = 1, \ldots, n$ and with equality for $d = n$. We use the same notation for any permutations of x and y.

We state the following basic fact about majorization without proof.

Theorem 12.1.3. *For two n-component real vectors x and y, it holds $x \prec y$ if and only if there exists a bistochastic matrix B such that $x_i = \sum_j B_{ij} y_j$.*

Now, as a direct consequence of the Birkhoff theorem, which states that a matrix is bistochastic if and only if it is a convex combination of permutation matrices, the concavity of the entropy implies the following.

Theorem 12.1.4. *Given two probability vectors p and q, if $p \prec q$, then $H(p) \geq H(q)$, in that case p is more disordered than q.*

12.1.2 *Characterization of LOCC transformations*

Theorem 12.1.5. *A bipartite state vector $|\psi\rangle$ can be transformed into $|\phi\rangle$ by LOCC transformations if and only if $\lambda_\psi \prec \lambda_\phi$, where λ_ψ, λ_ϕ are the vectors of Schmidt coefficients of $|\psi\rangle$ and $|\phi\rangle$, respectively.*

To prove the theorem, we need a preliminary result.

Definition 12.1.6. Given two Hermitian operators H, K, we say that $H \prec K$ if $\lambda(H) \prec \lambda(K)$, where $\lambda(H)$, $\lambda(K)$ are the vectors of their respective eigenvalues.

Lemma 12.1.7. *$H \prec K$ if and only if there exist unitary operators U_j and a probability distribution p_j such that*

$$H = \sum_j p_j U_j K U_j^\dagger. \tag{12.1}$$

Proof. Diagonalizing the operators, we have $H = V^\dagger \Lambda(H) V$, $K = T\Lambda(K)T^\dagger$, for suitable unitary operators V and U.

We now assume that Eq. (12.1) holds true. This is equivalent to the equation

$$\begin{aligned}\Lambda(H) &= \sum_j p_j \left(VU_jT\right)\Lambda(K)\left(T^\dagger U_j^\dagger V^\dagger\right)\\ &= \sum_j p_j \Omega_j \Lambda(K)\Omega_j^\dagger.\end{aligned}$$

Rewriting it in components,

$$\lambda_h(H) = \sum_j p_j \sum_k (\Omega_j)_{hk}\lambda_k(K)(\Omega_j^*)_{hk} \tag{12.2}$$

$$= \sum_k \sum_j p_j |(\Omega_j)_{hk}|^2 \lambda_k(K). \tag{12.3}$$

We can now note that the matrix $\sum_j p_j |(\Omega_j)_{hk}|^2$ is bistochastic, hence $\lambda(H) \prec \lambda(K)$.

Vice versa, let us suppose that $\lambda(H) \prec \lambda(K)$. It follows that there exists a bistochastic matrix B such that

$$\lambda_h(H) = \sum_k B_{hk}\lambda_k(K).$$

Using the Birkhoff theorem, we can always find a probability vector $(p_1, p_2, \ldots)$ and permutation matrices P_1, $P_2, \ldots$ such that $B = \sum_j p_j P_j$, then

$$\begin{aligned}\lambda_h(H) &= \sum_j p_j \sum_k (P_j)_{hk}\lambda_k(K)\\ &= \sum_j p_j \sum_k |(P_j)_{hk}|^2 \lambda_k(K).\end{aligned}$$

The lemma is finally proven by noting that the permutation matrix is also unitary, hence the latter equation is of the same form of Eqs. (12.2) and (12.3). □

Proof of Theorem 12.1.5. In the LOCC protocol, Alice makes a single measurement, with operators $\{M_j\}$, then Bob makes a local unitary. On Alice's side, the reduced density ρ_ψ is transformed to ρ_ϕ, and this should be obtained for all measurement results (j), that is,

$$M_j \rho_\psi M_j^\dagger = p_j \rho_\phi.$$

Making the polar decomposition, we have

$$M_j \sqrt{\rho_\psi} = \sqrt{p_j \rho_\phi}\, V_j,$$

for some unitary operator V_j. Multiplying by its adjoint equation, we get

$$\sqrt{\rho_\psi} M_j^\dagger M_j \sqrt{\rho_\psi} = p_j V_j^\dagger \rho_\phi V_j,$$

and summing over j (using the POVM completeness relation)

$$\rho_\psi = \sum_j p_j V_j^\dagger \rho_\phi V_j.$$

From Lemma 12.1, it follows that $\lambda_\psi \prec \lambda_\phi$.

Vice versa, assume that two states satisfy $\lambda_\psi \prec \lambda_\phi$, then according to the previous theorem, there exist unitaries U_j and probabilities p_j such that $\rho_\psi = \sum_j p_j U_j \rho_\phi U_j^\dagger$.

Let us first suppose that ρ_ψ is invertible, then we define the operators M_js as

$$M_j = \sqrt{p_j \rho_\phi}\, U_j^\dagger\, \rho_\psi^{-1/2}.$$

The operators $\{M_j\}$ define a POVM. In fact, they satisfy the completeness condition

$$\sum_j M_j^\dagger M_j = \rho_\psi^{-1/2}\, U_j p_j \rho_\phi U_j^\dagger\, \rho_\psi^{-1/2}$$

$$= \rho_\psi^{-1/2}\, \rho_\psi\, \rho_\psi^{-1/2} = I.$$

That is, if Alice performs the POVM with elements $\{M_j\}$, she realizes the transformation $\rho_\psi \to \rho_\phi$.

Finally, if ρ_ψ is not invertible, we can redefine the Hilbert space of Alice as the support of the operator ρ_ψ and then proceed as above (since an operator is always invertible on its support). □

12.1.3 *Entanglement monotones*

The pure states of a bipartite quantum system are then ordered according to the property that one state can be obtained from another state by LOCC operations.

Definition 12.1.8 (Entropy of entanglement). Given a bipartite pure quantum state $|\psi\rangle_{AB}\langle\psi|$, its entropy of entanglement is $S_e(|\psi\rangle_{AB}) := S(\rho_A) = S(\rho_B)$, where ρ_A and ρ_B are the reduced states.

It follows that $S_e(|\psi\rangle_{AB}) = H(\lambda)$, where λ is the vector of the Schmidt coefficients and H denotes the Shannon entropy.

We hence have that if $|\psi\rangle_{AB}$ can be transformed into $|\phi\rangle_{AB}$ by LOCC transformations, then $S_e(|\psi\rangle_{AB}) \geq S_e(|\phi\rangle_{AB})$.

Definition 12.1.9. An entanglement monotone is any function of a bipartite quantum state, with the following properties:

(1) $E(\rho_{AB}) \geq 0$.
(2) $E(\rho_{AB}) = 0$ if ρ_{AB} is separable.
(3) E is constant under local unitaries, and cannot increase under LOCC.
(4) Convexity: $E\left(\sum_k p_k \rho^k_{AB}\right) \leq \sum_k p_k E(\rho^k_{AB})$ (sometimes this property is not required).

Entanglement monotones are used as measures of the amount of entanglement.

The entropy of entanglement is an entanglement monotone on the set of pure states. More general examples are provided below.

Example. $E(\rho) := \inf_{\sigma \in \mathcal{S}} D(\rho, \sigma)$ where D is a distance-like quantity (actually, need not be a metric) and $\mathcal{S}$ is the set of separable states. In particular, D can be the relative entropy.

Example. $E(\rho) := \inf_{p_i|\psi_i\rangle} \sum_i p_i E(|\psi_i\rangle\langle\psi_i|)$ where $E(|\psi_i\rangle\langle\psi_i|)$ is an entanglement monotone on the set of pure states.

12.2 Entanglement distillation

Definition 12.2.1. Consider n copies of a bipartite state ρ_{AB}. Let m be the number of maximally entangled two-qubit states that can be distilled (with fidelity approaching 1 for $n \to \infty$) by means of LOCC transformations. Then, the asymptotic maximum rate m/n of distillation is called distillable entanglement,

$$D_e(\rho_{AB}) := \lim_{n\to\infty} \max_{\text{LOCC}} \frac{m}{n},$$

where the maximum is over all possible LOCC transformations.

The distillable entanglement is an entanglement monotone (properties 1–3 can be easily verified).

The distillable entanglement equals the number of maximally entangled states that can be distilled per copy of the given state.

For a bipartite state if up to $D_e(\rho_{AB})$ maximally entangled states can be distilled, we say that the ρ_{AB} has $D_e(\rho_{AB})$ *ebits* of distillable entanglement.

Theorem 12.2.2. *For bipartite state vectors $|\psi\rangle_{AB}$, the distillable entanglement is equal to the entanglement entropy, $D_e(|\psi\rangle_{AB}) = S_e(|\psi\rangle_{AB})$.*

Proof. Let us assume that Alice and Bob operate on qubits, the extension to Hilbert space of higher dimension is straightforward.

We first prove that $D_e(|\psi\rangle_{AB}) \leq S_e(|\psi\rangle_{AB})$. To do that, let us note that the entanglement entropy of $|\psi\rangle_{AB}^{\otimes n}$ is $S_e(|\psi\rangle_{AB}^{\otimes n}) = nS_e(|\psi\rangle_{AB})$, and that the entanglement entropy of m copies of a maximally entangled state vector is equal to $S(|\Phi^+\rangle^{\otimes m}) = m$. Since the entanglement entropy cannot increase under LOCC, we have that, $m/n \leq S_e(|\psi\rangle_{AB})$.

We now prove that $D_e(|\psi\rangle_{AB}) \geq S_e(|\psi\rangle_{AB})$. For doing that, we introduce an example of LOCC protocol with rate $m/n = S_e(|\psi\rangle_{AB})$.

The state vector under consideration is written in the Schmidt form, as $|\psi\rangle_{AB} = \sum_{x=0,1} \sqrt{p(x)}|x\rangle_{AB}|x\rangle_{AB}$, with $x = 0, 1$ for a two-qubit system. Alice and Bob have at their disposal n copies, $|\psi\rangle_{AB}^{\otimes n}$. We use the notion of typical sequences,

$$\begin{aligned}|\psi\rangle_{AB}^{\otimes n} &= \sum_{x_j=0,1} \sqrt{p(x_1)p(x_2)\cdots p(x_n)}|x_1, x_2 \cdots x_n\rangle_A|x_1, x_2 \cdots x_n\rangle_B \\ &\simeq \sum_{x_1x_2\cdots x_n \in T_X(n,\epsilon)} \sqrt{p(x_1)p(x_2)\cdots p(x_n)}|x_1, x_2 \cdots x_n\rangle_A \\ &\quad \otimes |x_1, x_2 \cdots x_n\rangle_B = |\psi_{typ}\rangle,\end{aligned}$$

where $\simeq$ means that the two state vectors cannot be distinguished asymptotically (their fidelity is bigger than $1 - \epsilon$). Alice can hence transform $|\psi\rangle_{AB}^{\otimes n}$ into $|\psi_{typ}\rangle$ by making a local projection into the subspace $span\{|x_1, x_2 \ldots x_n\rangle_A\}_{x_1x_2\ldots x_n \in T_X(n,\epsilon)}$. This projection will be successful with a probability approaching one in the limit of big n. The typical state vector $|\psi_{typ}\rangle$ is not normalized, but it can be normalized as $|\psi'_{typ}\rangle = |\psi_{typ}\rangle/\sqrt{\langle\psi_{typ}|\psi_{typ}\rangle}$. After the normalization, recalling that the number of typical sequences is asymptotically equal to $2^{nH(p)} = 2^{nS_e(|\psi\rangle_{AB})}$, we may note that

$$|\psi'_{typ}\rangle = \frac{1}{\sqrt{2^{nS_e(|\psi\rangle_{AB})}}} \sum_{x_1x_2\cdots x_n \in T_X(n,\epsilon)} |x_1, x_2 \cdots x_n\rangle_A|x_1, x_2 \cdots x_n\rangle_B.$$

Finally, note that this state is equivalent, up to local unitaries, to the product of $nS_e(|\psi\rangle_{AB})$ copies of a maximally entangled state vector of two qubits. □

It should be noted that the distillable entanglement of Definition 12.2.1 can be turned into a *capacity* (of distillable entanglement) notion, which results analogous to the quantum capacity.

Take a maximally entangled state vector on the space $\mathcal{H} \otimes \mathcal{H}$,

$$|\phi^+(\mathcal{H})\rangle = \frac{1}{\sqrt{d}} \sum_{i=1}^{d} |ii\rangle, \tag{12.4}$$

where $|i\rangle$ are basis vectors in $\mathcal{H}$ with $d = |\mathcal{H}|$. Given a state ρ, consider the sequence of operations $\{\mathcal{D}^n\}$ transforming the input state $\rho^{\otimes n}$ into the state τ_n acting on the Hilbert space $\mathcal{H}_n \otimes \mathcal{H}_n$, with $|\mathcal{H}_n| = d_n$, satisfying

$$F_n \equiv \langle \phi^+(\mathcal{H}_n)|\tau_n|\phi^+(\mathcal{H}_n)\rangle \overset{n\to\infty}{\longrightarrow} 1. \tag{12.5}$$

The asymptotic ratio attainable via the given protocol (sequence of operations) is then given by

$$D_{\{\mathcal{D}^n\}}(\rho) = \lim_{n\to\infty} \frac{\log d_n}{n}. \tag{12.6}$$

The *distillable entanglement capacity* $D(\rho)$ is defined by the supremum overall possible protocols

$$D(\rho) := \sup_{\{\mathcal{D}^n\}} D_{\{\mathcal{D}^n\}}(\rho). \tag{12.7}$$

After all, the quantum capacity of Section 11.4 is defined in the same way and simply follows by making the substitutions: $D \to Q$, $\rho \to \mathcal{N}^{\otimes n}(\rho)$ and $\mathcal{D} \to \mathcal{C}$ (where $\mathcal{N}$ is a quantum channel and $\mathcal{C}$ is a quantum error correction code). As a consequence, the distillable entanglement capacity can be expressed in terms of coherent information, namely $D = I_{\text{coh}}(\rho, \text{id})$.

12.3 Entanglement dilution

Definition 12.3.1. Given a number m of maximally entangled two-qubit states let n be the number of copies of a state ρ_{AB} that can be obtained (with fidelity approaching 1 in the limit of $n \to \infty$) by means of LOCC transformations. Then, the asymptotic minimum rate m/n is by definition the *entanglement of formation* of ρ_{AB}:

$$E_F(\rho_{AB}) := \lim_{n\to\infty} \min_{\text{LOCC}} \frac{m}{n}. \tag{12.8}$$

The entanglement of formation results in an entanglement monotone. It expresses the minimum number of maximally entangled two-qubit states that are needed to generate each copy of ρ_{AB}. For a bipartite state if a number of $E_F(\rho_{AB})$ maximally entangled states are required, we say that ρ_{AB} has $E_F(\rho_{AB})$ *ebits* of entanglement of formation.

Theorem 12.3.2. *For bipartite state vectors $|\psi\rangle_{AB}$, the entanglement of formation is equal to the entropy of entanglement.*

Proof. Let us assume that Alice and Bob operate on qubits, the extension to Hilbert space of higher dimension is straight forward.

First, we prove that $E_F(|\psi\rangle_{AB}) \geq S_e(|\psi\rangle_{AB})$: since the entanglement entropy cannot increase under LOCC, we have $m \geq nS_e(|\psi\rangle_{AB})$.

Then we prove that $E_F(|\psi\rangle_{AB}) \leq S_e(|\psi\rangle_{AB})$ by presenting an explicit LOCC protocol. For very large n, we have

$$\begin{aligned}|\psi\rangle_{AB}^{\otimes n} &= \sum_{x_j=0,1} \sqrt{p(x_1)p(x_2)\cdots p(x_n)}|x_1, x_2\cdots x_n\rangle_A|x_1, x_2\cdots x_n\rangle_B \\ &\simeq \sum_{x_1x_2\cdots x_n\in T_X(n,\epsilon)} \sqrt{p(x_1)p(x_2)\cdots p(x_n)}|x_1, x_2\cdots x_n\rangle_A \\ &\quad \times |x_1, x_2\cdots x_n\rangle_B = |\psi_{typ}\rangle.\end{aligned}$$

where, again, $\simeq$ means that the two state vectors cannot be distinguished asymptotically (their fidelity is bigger than $1-\epsilon$). As noticed in the proof of Theorem 12.2, such a state vector can be understood as $nS(|\psi\rangle_{AB})$ copies of a two-qubit maximally entangled state vector. Then, we note that if Alice and Bob initially share $m = nS(|\psi\rangle_{AB})$ maximally entangled state vectors of two qubits, they can make the teleportation of m arbitrary qubit states (teleportation only requires LOCC operations). So, Alice can prepare (locally) "both sides" of $|\psi'_{\text{typ}}\rangle = |\psi_{\text{typ}}\rangle/\sqrt{\langle\psi_{\text{typ}}|\psi_{\text{typ}}\rangle}$ (by making a local projection of $|\psi\rangle_{AB}^{\otimes n}$ into the subspace $span\{|x_1, x_2 \ldots x_n\rangle_A\}_{x_1x_2\cdots x_n\in T_X(n,\epsilon)}$) and teleport "one half" of it to Bob. In this way, Alice and Bob share $|\psi'_{typ}\rangle$, which asymptotically converges to $|\psi\rangle_{AB}^{\otimes n}$. The asymptotic dilution rate is hence $m/n = S_e(|\psi\rangle_{AB})$, implying $E_F(|\psi\rangle_{AB}) \leq S_e(|\psi\rangle_{AB})$. $\square$

Exercises

1. For $x, y \in \mathbb{R}^n$, show that it holds $x \prec y$ if and only if for all convex functions $f : \mathbb{R} \to \mathbb{R}$, $\sum_{i=1}^n f(x_i) \leq \sum_{i=1}^n f(y_i)$.
2. Show that Alice and Bob can convert $|\psi\rangle$ into $|\phi\rangle$ using only local operation without classical communication if and only if λ_ψ has non-zero entries identical to those of $\lambda_\phi \otimes \vec{p}$, where $\vec{p}$ is a probability vector.
3. Show that Alice and Bob cannot convert (with certainty) by LOCC the following two state vectors:
$$|\psi_1\rangle = \sqrt{0.4}|00\rangle + \sqrt{0.4}|11\rangle + \sqrt{0.1}|22\rangle + \sqrt{0.1}|33\rangle,$$
$$|\psi_2\rangle = \sqrt{0.5}|00\rangle + \sqrt{0.25}|11\rangle + \sqrt{0.25}|22\rangle,$$
that is $|\psi_1\rangle \not\leftrightarrow |\psi_2\rangle$. Then show that they can convert the two state vectors if they are supplied by a two-qubit state vector
$$|\phi\rangle = \sqrt{0.6}|44\rangle + \sqrt{0.4}|55\rangle.$$
That is, $|\psi_1\rangle|\phi\rangle \leftrightarrow |\psi_2\rangle|\phi\rangle$. This is the idea of *entanglement catalysis*, with $|\phi\rangle$ being the *catalyst*. Further, show that no transformation can be catalyzed by a maximally entangled state vector $|\phi\rangle = (1/\sqrt{d}) \sum_{i=1}^d |i_A\rangle|i_B\rangle$.
4. For two-qubit system, an entanglement monotone is the *concurrence* defined as
$$C(\rho) := \max\{0, \lambda_1 - \lambda_2 - \lambda_3 - \lambda_4\},$$
where $\lambda_1, \lambda_2, \lambda_3, \lambda_4$ are the singular values in decreasing order of $\sqrt{\rho}\sqrt{\theta\rho\theta}$, θ being the anti-unitary transformation (see appendix) acting as
$$\theta|\psi\rangle = (Y \otimes Y)|\psi\rangle^*,$$
with $|\psi\rangle^*$ the vector obtained from $|\psi\rangle$ by conjugation of components on the canonical basis.

Show for pure state that entanglement of formation can be expressed as

$$E_F(\rho) = H_2\left(\frac{1+\sqrt{1-C^2(\rho)}}{2}\right),$$

(the relation holds true for mixed states as well).

5. Show that for a depolarizing quantum channel with $p = 3/4$, a non-zero entanglement distillation rate $D(\rho)$ can be achieved though the quantum capacity is $Q = 0$.

Chapter 13

CRYPTOGRAPHY

The goal of cryptography is to make possible for two legitimate users (Alice and Bob) to exchange a message in a way not intelligible to a third malicious party (Eve, the eavesdropper). Here, we briefly present the most relevant classical cryptosystems and then introduce the quantum key distribution by analyzing the advantage it offers in terms of security.

13.1 Cryptosystems

Definition 13.1.1. A cryptosystem is determined by the following elements:

(i) a finite alphabet $\mathcal{A}$;
(ii) space of messages $\mathcal{M}$ on $\mathcal{A}$, i.e. $\mathcal{A}^n$;
(iii) space of keys $\mathcal{K} \subset \mathbb{N}$;
(iv) encryption function $E_k : \mathcal{A}^b \to \mathcal{A}^b$, with $k \in \mathcal{K}$ and $1 \leq b \leq n$;
(v) decryption function $D_{k'} : \mathcal{A}^b \to \mathcal{A}^b$ ($D_{k'} \equiv E_k^{-1}$) with $k' \in \mathcal{K}$ and $1 \leq b \leq n$.

If $k = k'$ (resp., $k \neq k'$), the cryptosystem is called symmetric (resp., asymmetric).

Example. A simple example is provided by the Caesar cipher where

- $\mathcal{A} = \{0, 1, \ldots, 25\}$,
- $\mathcal{M} = \mathcal{A}^n$,

- $\mathcal{K} = \{1, \ldots, 25\}$,
- $E_k : \mathcal{A} \to \mathcal{A}$ such that $E_k(m) = m + k \ (mod\ 26) = c$,
- $D_k : \mathcal{A} \to \mathcal{A}$ such that $D_k(c) = c - k \ (mod\ 26) = m$.

This cryptosystem has quite a small space for the key, hence can be easily broken by exhaustive search.

A more secure cryptosystem would involve not only cyclic permutation of the alphabet letters, but any possible permutation π, that is

- $\mathcal{K} = \{1, \ldots, |\mathcal{A}|! - 1\}$,
- $E_k : \mathcal{A} \to \mathcal{A}$ such that $E_k(m) = \pi_k(m) = c$,
- $D_k : \mathcal{A} \to \mathcal{A}$ such that $D_k(c) = \pi_k^{-1}(c) = m$.

However, by resorting to frequency analysis, this cryptosystem can also be broken. In fact, in any language, every letter has a characteristic frequency of occurrence, e.g. in English the most frequent letter is "E" with 12.31% use, then follows "T" with 9.59%, and then "A" with 8.05%, etc.

A remedy could be moving from *monoalphabetic substitution* to *poly-alphabetic substitution*, that is enciphering by using permutations on blocks of letters, not on single letters.

- $\mathcal{K} = \{1, \ldots, (|\mathcal{A}|! - 1)^b\}$,
- $E_k : \mathcal{A}^b \to \mathcal{A}^b$ such that $E_k(m) = \pi_k(m) \equiv \pi_{k_1}(m_1)\pi_{k_2}(m_2)\ldots \pi_{k_b}(m_b)$ with $m_i \in \mathcal{A}$, $k_i \in \{0, 1, \ldots, |\mathcal{A}|!\}$,
- $D_k : \mathcal{A}^b \to \mathcal{A}^b$ such that $D_k(c) = \pi_k^{-1}(c) = m$.

Unfortunately, this cryptosystem has also been shown to be breakable. It is enough to build up b messages by taking a letter every b letters of the original message so that each message results enciphered by a single permutation π_{k_i}. Then on each of the b messages, one has to apply the frequency analysis.

At this point, it seems natural to ask about the existence of an unbreakable cryptosystem!

13.2 The Vernam Cryptosystem

The Vernam cryptosystem dates back to 1917 and is characterized by

- $\mathcal{K} = \mathcal{M} = \mathcal{A}^n$,
- $E_k : \mathcal{A}^n \to \mathcal{A}^n$ such that $c = m + k$ (mod$|\mathcal{A}|$, letterwise),
- $k \in \mathcal{A}^n$ chosen in a random way, i.e. $\Pr(K = k) = \frac{1}{|\mathcal{A}|^n}$,
- $D_k : \mathcal{A}^n \to \mathcal{A}^n$ such that $m = c - k$ (mod$|\mathcal{A}|$, letterwise),
- the key is used only once (for this reason the cryptosystem is also known as *One-Time Pad* (OTP)).

Definition 13.2.1. A cryptosystem is perfect if given $c \in \mathcal{M}$, it holds $\Pr(E_k(m) = c) = \frac{1}{|\mathcal{A}|^n}$, $\forall m \in \mathcal{M}$, i.e. all plaintexts $m \in \mathcal{M}$ equally contribute to the ciphertext c.

Theorem 13.2.2. *One-Time Pad is a perfect cryptosystem.*

Proof. It immediately follows from the fact that $\Pr(m|c) = \Pr(m)$. As a consequence, it holds also $H(m|c) = H(m)$, which means that by knowing the ciphertext c we do not reduce our uncertainty about the plaintext m. □

Unfortunately, although being perfect, hence absolutely secure, One-Time Pad has several drawbacks:

- the key must be as long as the message and used only once, hence it needs a large amount of key material;
- the key must be truly random;
- the key must be shared by Alice and Bob beforehand.

The latter, also known as *key distribution problem*, is a serious problem that makes the OTP impractical for many applications.

13.3 Public-key Cryptography

A solution to the key distribution problem is provided by the public-key cryptosystem.

The Diffie and Hellman cryptosystem (1976) is a symmetric cryptosystem that reads as follows:

(i) Alice and Bob choose a large prime number N and an integer $g < N$;
(ii) N and g are made public;
(iii) Alice chooses a random integer x and sends to Bob, $u = g^x \pmod N$;
(iv) Bob chooses a random integer y and sends to Alice, $v = g^y \pmod N$;
(v) Alice computes $v^x \pmod N = g^{yx} \pmod N$;
(vi) Bob computes $u^y \pmod N = g^{xy} \pmod N$.

Alice and Bob end up with the same key $k = g^{xy} \pmod N$.

Eve, in order to learn k, should compute $x = \log_g u \pmod N$ and $y = \log_g v \pmod N$. Now the security of this protocol relies on the "difficulty" of computing these discrete logarithms. Here, "difficult" means computationally difficult, that is the number of computational steps needed to compute $x = \log_g u \pmod N$ scales in a non-polynomial way with respect to the length of the input $(\log N)$.

The RSA (Rivest, Shamir and Adleman, 1977) cryptosystem is another public cryptosystem, this one asymmetric.

(i) Alice chooses two large prime numbers p and q, and computes $N = p \times q$;
(ii) Alice randomly chooses an integer e coprime with $\phi(N) = (p-1)(q-1)$, and computes d such that $ed = 1 \,(\mathrm{mod}\, \phi(N))$ (i.e. $ed = 1 + j\phi(N)$);
(iii) Alice makes public N and e;
(iv) Bob represents a message (block of letters) by an integer m coprime with N;
(v) Bob computes $c = m^e \pmod N$ and sends it to Alice;
(vi) Alice receives c and computes $c^d \pmod N$, thus recovering m, in fact $c^d = m^{ed} = m^{1+j\phi(N)} = m \times \left(m^{\phi(N)}\right)^j \equiv m \pmod N$ where the last step follows from an Euler Theorem (1736) stating that: $gcd(m, N) = 1 \Rightarrow m^{\phi(N)} = 1 \pmod N$.

Note that the integer d at step (ii) can be efficiently computed by resorting to a variant of the Euler algorithm used to compute the *gcd*.

Eve, in order to learn m, should know d and in turn to get it she should factorize N in prime numbers $q \times p$. Then the security of this protocol relies on the "difficulty" of factoring large numbers.

The security of Diffie–Hellman and RSA protocols is called *conditional security* because it is conditional to the computation capabilities.

13.4 Quantum key distribution

Quantum information provides an alternative way to solve the key distribution problem. Moreover, this solution guarantees an *unconditional security*.

13.4.1 *BB84 protocol (Bennett and Brassard, 1984)*

1. Alice and Bob agree to associate a classical bit to the following state vectors of one qubit:

$$\begin{aligned}
&(\text{bit value } 0) \leftrightarrow |0\rangle,\\
&(\text{bit value } 1) \leftrightarrow |1\rangle,\\
&(\text{bit value } 0) \leftrightarrow |+\rangle = \frac{|0\rangle + |1\rangle}{\sqrt{2}},\\
&(\text{bit value } 1) \leftrightarrow |-\rangle = \frac{|0\rangle - |1\rangle}{\sqrt{2}}.
\end{aligned}$$

2. Alice then randomly prepares one of these four state vectors, each with probability 1/4. Recall that $|0\rangle$, $|1\rangle$ are the eigenstates of σ^Z, and $|-\rangle$, $|+\rangle$ are those of σ^X.
3. Alice sends the qubit to Bob.
4. Bob randomly decides to make a measurement of either σ^Z or σ^X, the result of the measurement is then associated to the bit values 0 or 1.
5. After repeating steps 1 to 4 n times, Alice and Bob have both a string of n binary numbers: those encoded by Alice and those

measured by Bob. The two strings may be different each time Alice encodes in one basis (e.g. the eigenstates of σ^Z) and Bob measures in the other basis (e.g. the eigenvalues of σ^X). Then, Alice and Bob publicly declare the basis on which they have prepared (Alice) and measured (Bob) the qubits. Note that they need to send n bits of information each.

6. After having exchanged this information, Alice and Bob select and discard the bits for which there is a mismatch of the basis. Hence, they end up with two shorter strings, whose bits correspond to equal choices of the basis, which are identical to each other (the length of these strings is asymptotically equal to $n/2$).

What does one gain by using this quantum protocol?

Let us analyze the simplest eavesdropping strategy known as *intercept and resend.* Eve takes the qubit from Alice, measures it (either in the σ^X or σ^Z basis with equal probability), then sends the resulting state vector to Bob. It obviously suffices to consider that Alice and Bob are using the same basis. Then, on the one hand, if Eve uses the same basis used by Alice, she can perfectly understand the bit value encoded by Alice (which will be the same measured by Bob) and this happens with probability $1/2$; on the other hand, if Eve uses a basis which is different from that used by Alice, her bit value is random, and the one got by Bob will also be random and the probability that Bob gets a different value from Alice will be $1/4$.

Now Alice and Bob extract a (long enough) part of their secret strings, and they publicly declare the bit values of the substrings. Without Eve, the two substrings should be equal to each other. If they find such a non-zero rate of bit mismatch ($1/4$ in this case), they can argue that someone is spying on them, then abort the protocol.

In summary, Alice and Bob cannot prevent Eve's intervention, but can always outwit her.

13.4.2 *E91 protocol (Ekert, 1991)*

This protocol uses entanglement as a resource.

1. Alice and Bob share entanglement through a maximally entangled state vector (actually, n copies of it), e.g.

$$|\Phi^+\rangle = \frac{|00\rangle + |11\rangle}{\sqrt{2}} = \frac{|{+}{+}\rangle + |{-}{-}\rangle}{\sqrt{2}}.$$

2. Alice measures either σ^Z or σ^X, and Bob does the same. They collect the binary strings with their measurement outcomes.
3. Alice and Bob publicly declare the basis they have used, and proceed as in BB84.

This protocol is equivalent to BB84. To show this, it would be enough to consider the two entangled qubits in the state vector $|\Phi^+\rangle$ originally in the hands of Alice who performs measurement of either σ^Z or σ^X observable chosen randomly. Hence, she will obtain the state vectors $|0\rangle$, $|1\rangle$, $|+\rangle$, $|-\rangle$ each with probability 1/4 that she will send to Bob who will randomly measure σ^Z or σ^X.

13.5 Quantum security

Of course, aborting the quantum key distribution protocol every time Alice and Bob find a non-zero error rate in their bit strings is even more unpractical than OTP. In fact, errors could also be due to other effects rather than the intervention of an eavesdropper. So it is desirable to have a higher than zero tolerable bound for the error rate.

First of all, note that from a quantum bit error rate (QBER) q' of the testing sample (part of the key to sacrifice), Alice and Bob can ascertain the QBER q of the untested sample with a high confidence assuming $q = q'$. The confidence level is established by the so-called *random sampling* according to which, for m tested bits and any $\epsilon > 0$, we have

$$\Pr\{\, q > (q' + \epsilon)\} < \exp[-O(\epsilon^2 m)].$$

Now, given a QBER q, we can immediately say that the quantum channel is analogous to a binary symmetric channel, so that the Alice and Bob conditional entropies amount to

$$H(A|B) = H(B|A) = H(q). \tag{13.1}$$

The security of BB84 (as well as E91) QKD protocol relies on an entropic version of the uncertainty relation stated by the following theorem.

Theorem 13.5.1. *For any density operator ρ_{AB} on $\mathcal{H}_A \otimes \mathcal{H}_B$ and pair of observables O_1 and O_2 in $\mathcal{H}_A$, it holds*

$$S(O_1|B) + S(O_2|B) \geq 2\log\frac{1}{c} + S(A|B), \tag{13.2}$$

where $S(O_i|B)$ is computed on the state

$$\sum_j (|o_i^j\rangle\langle o_i^j| \otimes I)\rho_{AB}(|o_i^j\rangle\langle o_i^j| \otimes I),$$

$|o_1^j\rangle$, $|o_2^j\rangle$ being the eigenvectors of O_1 and O_2, respectively. Furthermore, $c := \max_{j,k} |\langle o_1^j|o_2^k\rangle|$.

Proof (sketch). The term $S(A|B)$ appearing on the right hand side quantifies the amount of entanglement between the systems A and B. Note that if A and B are maximally entangled, then $S(A|B) = -\log d$. Since $2\log\frac{1}{c}$ cannot exceed $\log d$, the bound of Theorem 13.5.1 reduces to $S(O_1|B) + S(O_2|B) \geq 0$, which is trivial, that is Bob can guess both O_1 and O_2 perfectly. On the other hand, if A and B are not entangled, then $S(A|B) \geq 0$. Since $S(O_i|B) \leq S(O_i)$ for all states, the bound (13.2) reduces to

$$H(O_1) + H(O_2) \geq 2\log\frac{1}{c}, \tag{13.3}$$

where $H(O_i)$ denotes the Shannon entropy of the probability distribution of the outcomes when O_i is measured. For the sake of simplicity, we will simply provide the proof for this latter case.

Let $P^{(i)}$ denote the probability distribution over a set of N possible outcomes for O_i and define

$$M_r\left(P^{(i)}\right) := \left[\sum_j (P_j^{(i)})^{1+r}\right]^{1/r}, \tag{13.4}$$

with $-1 \leq r \leq +\infty$. Such a quantity can be regarded as measuring the average peakedness of $P^{(i)}$. Also notice that $M_0(P^{(i)}) = \exp[-H(P^{(i)})]$.

We are now going to exploit an inequality which is a consequence of a famous theorem on bilinear forms (Riesz Theorem). Namely, given a sequence of complex numbers $x \equiv (x_1, \ldots, x_N)$ and a matrix $\{T_{jk}\}$ representing a linear transformation T acting as $(Tx)_j = \sum_k T_{jk} x_k$ obeying the condition $\sum_j |(Tx)_j|^2 = \sum_j |x_k|^2$ for all x and letting $c := \max_{jk} |T_{jk}|$, we have

$$c^{1/a'}\left[\sum_j |(Tx)_j|^{a'}\right]^{1/a'} \leq c^{1/a}\left[\sum_j |x_k|^a\right]^{1/a}, \tag{13.5}$$

with $1 \leq a \leq 2$ and $\frac{1}{a} + \frac{1}{a'} = 1$.

Suppose that the state $\rho^A = \mathrm{Tr}_B(\rho^{AB})$ is pure, i.e. equal to $|\psi\rangle\langle\psi|$. Then we can take

$$\begin{aligned} x_k &= \langle o_1^k|\psi\rangle, \\ T_{jk} &= \langle o_2^j|o_1^k\rangle, \\ (Tx)_j &= \langle o_2^j|\psi\rangle, \end{aligned}$$

and putting $a = 2(1+r)$, $a' = 2(1+s)$, we obtain

$$M_r(P^{(1)}) M_s(P^{(2)}) \leq c^2, \tag{13.6}$$

for $r = -s/(1+2s)$, $s \geq 0$. Furthermore, taking the limit $s \to 0$ in the above expression, we arrive at (13.3).

For a general mixed state $\rho^A = \sum_n a_n |\psi^{(n)}\rangle\langle\psi^{(n)}|$, we may write the probability distributions corresponding to the outcomes of the observables O_1, O_2 as

$$\overline{P^{(1)}}_j = \sum_n a_n P_j^{(1,n)}, \qquad \overline{P^{(2)}}_j = \sum_n a_n P_j^{(2,n)}.$$

Then the inequality (13.6) for $s = 0$ immediately follows from the fact that the Shannon entropy $H(P) = -\log M_0(P)$ is a concave function of P. □

Before analyzing the consequence of Theorem 13.5.1, we need to put forward a lower bound on the distillable key rate.

13.5.1 *Privacy and coherent information*

Quite generally, we can associate a quantum system E to Eve, and assume that Eve's system and Alice's system are coupled by a unitary transformation U_{AE}. Then Eve measures the output of her system. In this way, the received state by Bob is

$$\rho_B = \mathcal{N}(\rho_A) = \mathrm{Tr}_E[U_{AE}(\rho_A \otimes |0\rangle_E\langle 0|)U_{AE}^\dagger], \tag{13.7}$$

where $|0\rangle_E\langle 0|$ is the initial state of Eve's system, and ρ_A is the input state from Alice. Similarly, the state received by Eve is

$$\rho_E = \widetilde{\mathcal{N}}(\rho_A) = \mathrm{Tr}_B[U_{AE}(\rho_A \otimes |0\rangle_E\langle 0|)U_{AE}^\dagger]. \tag{13.8}$$

Note that $\widetilde{\mathcal{N}}$ is the complementary channel of $\mathcal{N}$ (see Definition 6.3.4).

If Alice encodes classical information via an ensemble of quantum states $\{p_i, \rho_i\}$, we know that the maximal information that Bob can retrieve is given by the Holevo function

$$\chi(\{p_i, \mathcal{N}(\rho_i)\}) = S\left(\sum_i p_i \mathcal{N}(\rho_i)\right) - \sum_i p_i S[\mathcal{N}(\rho_i)]. \tag{13.9}$$

Similarly, the maximum information that Eve can steal is given by

$$\chi(\{p_i, \widetilde{\mathcal{N}}(\rho_i)\}) = S\left(\sum_i p_i \widetilde{\mathcal{N}}(\rho_i)\right) - \sum_i p_i S[\widetilde{\mathcal{N}}(\rho_i)]. \tag{13.10}$$

The maximum rate of secret classical information shared by Alice and Bob is hence given by maximum of the difference

$$\max_{\{p_i,\rho_i\}} [\chi(\{p_i, \mathcal{N}(\rho_i)\}) - \chi(\{p_i, \widetilde{\mathcal{N}}(\rho_i)\})] = P_1 . \tag{13.11}$$

This quantity is the (single letter version of) *private capacity* of the channel $\mathcal{N}$.

A lower bound on it can be obtained by restricting the maximization over pure input states

$$P_1 \geq \max_{\{p_i,|\psi_i\rangle\}} [\chi(\{p_i, \mathcal{N}(|\psi_i\rangle\langle\psi_i|)\}) - \chi(\{p_i, \widetilde{\mathcal{N}}(|\psi_i\rangle\langle\psi_i|)\})] . \tag{13.12}$$

The quantity in the square brackets is equal to the following coherent information:

$$\begin{aligned}
&\chi(\{p_i, \mathcal{N}(|\psi_i\rangle\langle\psi_i|)\}) - \chi(\{p_i, \widetilde{\mathcal{N}}(|\psi_i\rangle\langle\psi_i|)\}) \\
&\quad = S\left(\sum_i p_i \mathcal{N}(|\psi_i\rangle\langle\psi_i|)\right) - S\left(\sum_i p_i \widetilde{\mathcal{N}}(|\psi_i\rangle\langle\psi_i|)\right) \\
&\qquad - \sum_i p_i \left\{S[\mathcal{N}(|\psi_i\rangle\langle\psi_i|)] - S[\widetilde{\mathcal{N}}(|\psi_i\rangle\langle\psi_i|)]\right\} \\
&\quad = S\left(\sum_i p_i \mathcal{N}(|\psi_i\rangle\langle\psi_i|)\right) - S\left(\sum_i p_i \widetilde{\mathcal{N}}(|\psi_i\rangle\langle\psi_i|)\right) \\
&\quad = I_{coh}\left(\sum_i p_i |\psi_i\rangle\langle\psi_i|, \mathcal{N}\right),
\end{aligned}$$

where we have used the fact that $S\left[\mathcal{N}(|\psi_i\rangle\langle\psi_i|)\right] = S\left[\widetilde{\mathcal{N}}(|\psi_i\rangle\langle\psi_i|)\right]$. We hence have that the private capacity (13.11) is bigger than or equal to the quantum capacity (11.43),

$$P_1 \geq Q_1 . \tag{13.13}$$

The security of a QKD protocol can now be analyzed by assuming the worst case scenario where the eavesdropper holds the system E of a purification $|\Psi\rangle_{ABE}$ of ρ_{AB}, the state Alice and Bob share. This is because every other extension of ρ_{AB} can be clearly obtained from the purification by a quantum operation acting on E. Let us then consider Alice's and Bob's measurements chosen at random, with Alice's possible choices denoted by O_1 and O_2 and Bob's by O'_1 and O'_2.

We expect that they can generate a secure key if their measurement outcomes are sufficiently well correlated. To show this, let us consider the lower bound (13.12) on the amount of keys Alice and Bob can extract per state (key rate $\mathtt{K}$) that can be rewritten as

$$\begin{aligned} \mathtt{K} &\geq S(B) - S(E) = S(O_1, E) - S(E) - S(O_1, E) + S(B) \\ &= S(O_1|E) - S(O_1|B). \end{aligned} \tag{13.14}$$

with the entropic quantities to be computed on the state

$$\sum_j (|o_1^j\rangle\langle o_1^j| \otimes I \otimes I)\rho_{ABE}(|o_1^j\rangle\langle o_1^j| \otimes I \otimes I), \tag{13.15}$$

where $\rho_{ABE} = |\Psi\rangle_{ABE}\langle\Psi|$.

The equality in (13.14) has been obtained by noting that $S(O_1, E) = S(O_1, B)$. In fact, the state (13.15) can be rewritten as

$$\sum_j p_j |o_1^j\rangle\langle o_1^j| \otimes \rho_{BE}^j, \tag{13.16}$$

with

$$p_j := \mathrm{Tr}_{BE}\left[\langle o_1^j|\Psi\rangle_{ABE}\langle\Psi|o_1^j\rangle\right],$$

$$\rho_{BE}^j := \frac{1}{p_j}\langle o_1^j|\Psi\rangle_{ABE}\langle\Psi|o_1^j\rangle,$$

and therefore,

$$\begin{aligned} S(O_1, E) &= H(\vec{p}_j) + \sum_j p_j S\left(\mathrm{Tr}_B \rho_{BE}^j\right) \\ &= H(\vec{p}_j) + \sum_j p_j S\left(\mathrm{Tr}_E \rho_{BE}^j\right) = S(O_1, B), \end{aligned}$$

where ρ_{BE}^j is pure.

Now, we reformulate the inequality of Theorem 13.5.1 as

$$S(O_1|E) + S(O_2|B) \geq 2\log\frac{1}{c}. \tag{13.17}$$

In fact, the inequality of Theorem 13.5.1 also reads $S(O_1, B) + S(O_2, B) \geq 2\log\frac{1}{c} + S(A, B) + S(B)$. Since these quantities have to be evaluated on a pure state ρ_{ABE}, besides $S(O_1, B) = S(O_1, E)$ as shown above, we also have $S(A, B) = S(E)$. This yields $S(O_1, E) + S(O_2, B) \geq 2\log\frac{1}{c} + S(E) + S(B)$, which is equivalent to Eq. (13.17).

Together, Eqs. (13.14) and (13.17) imply $\mathtt{K} \geq 2\log\frac{1}{c} - S(O_1|B) - S(O_2|B)$. Furthermore, using the property that measurements cannot decrease entropy, we get

$$\mathtt{K} \geq 2\log\frac{1}{c} - S(O_1|O_1') - S(O_2|O_2'). \tag{13.18}$$

The entropies at the right hand side can be directly bounded by observable quantities such as the error rate q. Assuming qubit systems and $S(O_1|O_1') = S(O_2|O_2')$ by symmetry, we get, taking into account (13.1),

$$\mathtt{K} \geq 1 - 2H_2(q). \tag{13.19}$$

It then follows that $\mathtt{K} > 0$, provided $q < 11\%$.

Actually, standard information-theoretic arguments, like the above, are usually restricted to situations where the state ρ^N over N runs of the protocol is i.i.d., i.e. $\rho^N = \sigma^{\otimes N}$. This, however, is only guaranteed for *collective attacks*, where the adversary is bound to apply the same operation separately to each of the particles sent over the channel.

As an adversary might tamper with the particles sent over the (insecure) quantum channel, the joint state ρ^N of the N particle pairs held by Alice and Bob after the distribution phase is generally (almost) arbitrary. Hence, to prove security of the scheme against general attacks, one has to show that the distillation phase works correctly, whatever the state ρ^N is. Because the space of possible states ρ^N is exponentially large in N, this analysis is non-trivial.

Definition 13.5.2. The state ρ^N of an N-partite system is called symmetric if it is invariant under permutations of its subsystems, i.e. formally $\pi\rho^N\pi^\dagger = \rho^N$, where π is an arbitrary permutation.

This is equivalent to saying that the order in which the subsystems are represented mathematically is independent of their physical properties. Note that any i.i.d. state is symmetric, whereas the opposite implication does not generally hold. To make this more precise, consider the following.

Definition 13.5.3. Let ρ^n be a state on an n-partite quantum system as well as σ a state on a single subsystem. Then ρ^n is called $\binom{n}{m}$*-i.i.d. (with prototype* σ*)* if it has the form $\sigma^{\otimes m} \otimes \tilde{\rho}^{n-m}$, up to permutations of the subsystems, where $\tilde{\rho}^{n-m}$ is an arbitrary state on $n - m$ subsystems.

For $m = n$, we retrieve the standard notion of i.i.d. states.

Theorem 13.5.4 (Global representation theorem). *Any n-partite part ρ^n of an N-partite symmetric state ρ^N is approximated by a probabilistic mixture of states ρ^n_σ parameterizsed by σ, where each ρ^n_σ is contained in the space spanned by $\binom{n}{n-r}$-i.i.d. states with prototype σ, for $r \ll n$. The error of the approximation (quantified by trace distance) is upper bounded by $3\exp\left[-r\frac{N-n}{N} + d\ln(N-n)\right]$, where d is the dimension of the subsystems, i.e. the decrease is exponential in r.*

A typical choice for the above parameters is $n{:=}N - N^\alpha$ and $r := N^\alpha$, where $\frac{1}{2} < \alpha < 1$.

Roughly speaking, the global representation theorem then says that a symmetric state ρ^N can be seen as a mixture of i.i.d. states, as long as we ignore N^α subsystems and, additionally, tolerate deviations in at most N^α of the subsystems (note that N^α is only sublinear in N and the error ϵ decreases exponentially fast in N).

The proof of the *global representation theorem* goes beyond the scope of this book. Let us simply analyze its application in quantum cryptography.

Using the *global representation theorem*, it can be shown that security of a QKD scheme against *collective attacks* implies security against arbitrary attacks, namely *coherent attacks* where no restriction is imposed on the adversary. The argument is based on two observations:

(i) The security of the distillation phase only depends on *robust* properties of the state ρ^N of the N particle pairs held by Alice and Bob after the distribution phase, i.e. security is not affected by alterations of a small number of subsystems.

(ii) If Alice and Bob both reorder their particles according to a common randomly chosen permutation, then the resulting state of the particle pairs (averaged over all possible permutations) is *symmetric*.

Now, given a QKD scheme which is provably secure against collective attacks, observation (i) implies that the same scheme is secure whenever the state ρ^N has some n-partite part which is $\binom{n}{n-r}$-i.i.d., where $N - n \ll N$ and $r \ll n$. Hence, by global representation theorem, it suffices to verify that ρ^N is symmetric, which is the case because of observation (ii).

Thus, the following result holds true: If a QKD scheme is secure against collective attacks then the same scheme, equipped with an additional randomized permutation step inserted after the distribution phase is secure against any attacks, allowed by the laws of quantum physics (actually, inserting such a symmetrization step is only necessary if the scheme is not symmetric; many schemes are already symmetric by construction).

Exercises

1. Suppose that Alice would like to send Bob the message *I love you.* They agree to use RSA cryptosystem based on primes $p = 7$ and $q = 11$. Alice (resp., Bob) then chooses 2 (resp., 3) as private key. What are the public keys? In which way does Alice encrypt her message?

2. A simplified version of BB84 protocol can be obtained by using only two (non-orthogonal) states. Alice prepares one random bit a and depending on the result she sends to Bob a qubit in the state vector

$$\begin{array}{ll} |0\rangle & \text{if } a = 0, \\ \frac{|0\rangle + |1\rangle}{\sqrt{2}} & \text{if } a = 1. \end{array}$$

Depending on a random classical bit a' which he generates, Bob measures the qubit he receives in either the σ^Z basis (if $a' = 0$) or in the σ^X basis (if $a' = 1$). He then obtains the result b (0 or 1) corresponding to the $+1$ and -1 eigenvalues of σ^X and σ^Z. Bob publicly announces b, but keeps a' secret. Then, Alice and Bob keep only those pairs $\{a, a'\}$ for which $b = 1$. Only if $a' = 1 \oplus a$ will Bob obtain $b = 1$, so the final key is a for Alice and $1 \oplus a'$ for Bob.

Define a rate to establish the "raw" key as

$$\mathrm{K}_{raw} := \frac{\text{number of usable bits}}{\text{number of transmitted qubits} + \text{number of transmitted bits}}$$

and compare its value for BB84 and B92 protocols.

3. Show that if Alice and Bob share a state having fidelity $(1 - 2^{-s})$ with $|\Phi^+\rangle^{\otimes m}$, where $|\Phi^+ = (|00\rangle + |11\rangle)/\sqrt{2}$, then Eve's mutual information with the key is at most $\left(2^{-c} + 2^{O(-2s)}\right)$, where $c{:=}s - \log\left(2m + s + \frac{1}{\ln 2}\right)$.
4. Consider individual attacks by Pauli Cloning Machine (see Exercise 1, Chapter 6) able to copy $\{|0\rangle, |1\rangle\}$ and $\{|+\rangle, |-\rangle\}$ bases equally well and show that in such a case the QBER threshold is $\frac{\sqrt{2}-1}{2\sqrt{2}}$.
5. Given an N dimensional Hilbert space, an ensemble $\mathcal{E} \equiv \{p_i, \rho_i\}$ on it and two random variables B, E corresponding to the outcomes of two observables on it, derive the following entropic uncertainty relation

$$I(B : \mathcal{E}) + I(E : \mathcal{E}) \leq 2\log(Nc),$$

where $c = \max_{jk} |\langle b_j | e_k \rangle|$ with $|b_j\rangle$, $|e_k\rangle$ the eigenstates of the two observables.

Then, use this result together with the classical result (known as Csizar and Körner theorem) that for a given joint probability distribution $P_{A,B,E}$, Alice and Bob can establish a secret key if and only if $I(A : B) > I(A : E)$, to derive the threshold QBER of 11%.

APPENDIX: FUNDAMENTALS OF LINEAR ALGEBRA

In this appendix, the fundamental notions of linear algebra are revisited by employing the Dirac formalism.

A.1 Hilbert spaces

Definition A.1.1. A Hilbert space $\mathcal{H}$ is a vector space over the complex field $\mathbb{C}$ supplied with a positive definite scalar (inner) product such that it results complete in the norm induced by the scalar product.

Throughout this book, only finite dimensional Hilbert spaces are considered and each is isomorphic to $\mathbb{C}^n$. There the vectors are n-tuples of complex numbers that can be simply represented as column vectors. In the Dirac notation, the *ket* $|\psi\rangle$ indicates one such vector,

$$|\psi\rangle = \begin{pmatrix} \psi_1 \\ \psi_2 \\ \vdots \\ \psi_n \end{pmatrix}, \qquad \psi_j \in \mathbb{C}. \tag{A.1}$$

The scalar product in $\mathbb{C}^n$ between two vectors

$$|\phi\rangle = \begin{pmatrix} \phi_1 \\ \phi_2 \\ \vdots \\ \phi_n \end{pmatrix}, \qquad |\psi\rangle = \begin{pmatrix} \psi_1 \\ \psi_2 \\ \vdots \\ \psi_n \end{pmatrix}, \tag{A.2}$$

is defined as $\sum_{j=1}^{n} \phi_j^* \psi_j \in \mathbb{C}$. By Dirac notation, it is denoted as $\langle\phi|\psi\rangle$, $\langle\phi|$ being the *bra*, i.e. the dual, of $|\phi\rangle$.

Thus, the bra $\langle\phi|$ can be understood as a row vector (actually the transpose and conjugate of the ket $|\phi\rangle$), so that $\langle\phi|\psi\rangle$ becomes a row by column product.

Recall that the scalar product satisfies the following properties:

(i) $\langle\psi|\psi\rangle \geq 0$, $\forall|\psi\rangle \in \mathcal{H}$, with equality if and only if $|\psi\rangle = 0$ (null vector),
(ii) $\langle\phi|\left(a|\psi_1\rangle + b|\psi_2\rangle\right) = a\langle\phi|\psi_1\rangle + b\langle\phi|\psi_2\rangle$, $\forall|\phi\rangle, |\psi_1\rangle, |\psi_2\rangle \in \mathcal{H}$ and $a, b \in \mathbb{C}$,
(iii) $\langle\phi|\psi\rangle = \langle\psi|\phi\rangle^*$.

The norm induced by the scalar is

$$\||\psi\rangle\| = \sqrt{\langle\psi|\psi\rangle}. \tag{A.3}$$

Due to the finite dimensionality of $\mathbb{C}^n$, its completeness with respect to this norm is guaranteed by the completeness of $\mathbb{C}$. Namely each sequence of vectors $\{|v^{(k)}\rangle\}_k$ satisfying the Cauchy condition, $\forall\epsilon > 0$, $\exists K$ such that $\||v^{(k)}\rangle - |v^{(\ell)}\rangle\| < \epsilon$ when $k, \ell > K$, converge within $\mathbb{C}^n$.

A set of vectors $\{|i\rangle\}_i$ for which $\langle i|j\rangle = \delta_{ij}$ is clearly orthonormal.

An orthonormal basis $\{|e_i\rangle\}_{i=1}^n$ is a set of orthonormal vectors that span the entire space $\mathcal{H}$. The canonical basis is

$$|e_1\rangle = \begin{pmatrix} 1 \\ 0 \\ \vdots \\ 0 \end{pmatrix}, \quad |e_2\rangle = \begin{pmatrix} 0 \\ 1 \\ \vdots \\ 0 \end{pmatrix}, \quad \ldots, \quad |e_n\rangle = \begin{pmatrix} 0 \\ 0 \\ \vdots \\ 1 \end{pmatrix}. \tag{A.4}$$

Each subspace $\mathcal{K}$ of $\mathcal{H}$ is itself a Hilbert space and moreover it has an orthogonal complement $\mathcal{K}^\perp$, i.e. $\mathcal{H} = \mathcal{K} \oplus \mathcal{K}^\perp$.

A.2 Operators on Hilbert spaces

Definition A.2.1. A linear operator A is a map $A : \mathcal{H} \to \mathcal{H}$ such that[1]

$$A\left(a|\psi_1\rangle + b|\psi_2\rangle\right) = aA|\psi_1\rangle + bA|\psi_2\rangle,$$

for all $|\psi_1\rangle, |\psi_2\rangle \in \mathcal{H}$ and for all $a, b \in \mathbb{C}$.

We denote by $\mathfrak{L}(\mathcal{H})$ the set of linear operators $A : \mathcal{H} \to \mathcal{H}$.

Given two vectors $|v\rangle, |w\rangle \in \mathcal{H}$ beside their inner product, we can also consider their *outer product* $|v\rangle\langle w|$, which results a linear operator. In fact,

$$\left(|v\rangle\langle w|\right)|\psi\rangle = \left(\langle w|\psi\rangle\right)|v\rangle.$$

Let $\{|e_i\rangle\}_{i=1}^n$ be an orthonormal basis of $\mathcal{H}$, so that given $|v\rangle \in \mathcal{H}$, we can write $|v\rangle = \sum_i v_i|e_i\rangle$ with $v_i = \langle e_i|v\rangle \in \mathbb{C}$, then it holds

$$\left(\sum_{i=1}^n |e_i\rangle\langle e_i|\right)|v\rangle = \sum_{i=1}^n v_i|e_i\rangle, \quad \forall |v\rangle \in \mathcal{H},$$

which implies the following *completeness relation*:

$$\sum_{i=1}^n |e_i\rangle\langle e_i| = I. \tag{A.5}$$

As a consequence, given $A : \mathcal{H} \to \mathcal{H}$ linear, it holds

$$\begin{aligned} A &= \sum_{i=1}^n |e_i\rangle\langle e_i|A\sum_{j=1}^n |e_j\rangle\langle e_j| \\ &= \sum_{i,j=1}^n \langle e_i|A|e_j\rangle|e_i\rangle\langle e_j| \end{aligned} \tag{A.6}$$

with $A_{ij} := \langle e_i|A|e_j\rangle$ the entries of an $n \times n$ matrix representing A in the bases $\{|e_i\rangle\}_{i=1}^n$.

---

[1] All the notions can be straightforwardly extended to operators between different spaces.

The projector onto the subspace $\mathcal{K} \subset \mathcal{H}$ spanned by $\{|e_i\rangle\}_{i=1}^{k}$, $k < n$ is given by

$$P_{\mathcal{K}} = \sum_{i=1}^{k} |e_i\rangle\langle e_i|, \tag{A.7}$$

and it satisfies $P_{\mathcal{K}}^2 = P_{\mathcal{K}}$.

Definition A.2.2. Given $A : \mathcal{H} \to \mathcal{H}$, its adjoint operator $A^\dagger : \mathcal{H} \to \mathcal{H}$ is defined by means of

$$\langle v|Aw\rangle = \langle A^\dagger v|w\rangle, \quad \forall |v\rangle, |w\rangle \in \mathcal{H},$$

where $|Aw\rangle := A|w\rangle$ and $|A^\dagger v\rangle = A^\dagger|v\rangle$.

It results that

$$\langle v|Aw\rangle = \langle w|A^\dagger v\rangle^*$$

$$\langle v|Aw\rangle^* = \langle w|A^\dagger v\rangle$$

$$\langle v|A|w\rangle^* = \langle w|A^\dagger|v\rangle$$

and using an orthonormal basis $\{|e_i\rangle\}_{i=1}^{n}$

$$\langle e_i|A\langle e_j|^* = \langle e_j|A^\dagger|e_i\rangle,$$

i.e. $A_{ij}^* = A_{ji}$ meaning that the adjoint of A is represented by the transpose and conjugate matrix representing A. It holds also $|v\rangle^\dagger = \langle v|$.

Definition A.2.3. An operator $U : \mathcal{H} \to \mathcal{H}$ such that $U^\dagger U = I$ is called isometry. If additionally it holds also $UU^\dagger = I$, it is called unitary.

Note that isometries preserve the scalar product, hence the norm. That is, the scalar product of $U|v\rangle$ with $U|w\rangle$ equals $\langle v|u\rangle$ for all $|v\rangle, |w\rangle \in \mathcal{H}$. The same holds in particular for unitaries. If instead for the operator $U : \mathcal{H} \to \mathcal{H}$ it happens that the scalar product of $U|v\rangle$ with $U|w\rangle$ equals $\langle v|u\rangle^*$ for all $|v\rangle, |w\rangle \in \mathcal{H}$, the operator is said to be *antiunitary*.

Definition A.2.4. An operator $A : \mathcal{H} \to \mathcal{H}$ such that $A^\dagger = A$ is called self-adjoint or Hermitian (it is anti-Hermitian in the case $A^\dagger = -A$). If it satisfies $AA^\dagger = A^\dagger A$, the operator is said to be normal.

Projection operators are always Hermitian.

Definition A.2.5. An operator $A : \mathcal{H} \to \mathcal{H}$ such that $\langle v|A|v\rangle \geq 0$, $\forall |v\rangle \in \mathcal{H}$ is said to be positive.

A.2.1 *Spectral properties of operators*

Definition A.2.6. An eigenvector (eigenstate) of a linear operator $A : \mathcal{H} \to \mathcal{H}$ is a (non-null) vector $|i\rangle$ such that $A|i\rangle = a_i|i\rangle$, where $a_i \in \mathbb{C}$ is the corresponding eigenvalue.

The rank of an operator $A : \mathcal{H} \to \mathcal{H}$, denoted as rank$(A)$, is the dimension of the image of A. As such, it is equivalent to the number of its distinct and non-zero eigenvalues.

Theorem A.2.7. *Given $A : \mathcal{H} \to \mathcal{H}$ normal, if $\{a_i, |i\rangle\}_i$ is the set of its eigenvalues–eigenvectors, then $\{a_i^*, |i\rangle\}_i$ is the set of eigenvalues–eigenvectors of $A^\dagger$.*

As a corollary, we have that Hermitian operators have real eigenvalues and unitary operators have unimodular eigenvalues. As a consequence, positive operators are also Hermitian.

Theorem A.2.8 (Spectral Theorem). *Given $A : \mathcal{H} \to \mathcal{H}$ normal, there exists an orthonormal basis given by its eigenvectors, in which its matrix representation is diagonal. In other words, there exists a unitary operator $U : \mathcal{H} \to \mathcal{H}$ such that $A = UA_dU^\dagger$, with $A_d =$ diag$\{a_1, a_2, \ldots, a_n\}$, and a_j, $j = 1, \ldots, n$ being the eigenvalues of A.*

In summary, given $A : \mathcal{H} \to \mathcal{H}$ normal and $\{a_i, |i\rangle\}_i$ its eigenvalues and eigenvectors, we can express A through its *spectral decomposition*

$$A = \sum_i a_i|i\rangle\langle i| = \sum_\alpha \alpha P_\alpha, \tag{A.8}$$

where P_α is the projector onto the subspace of eigenvectors with eigenvalue α. It satisfies the relation

$$P_\alpha P_\beta = \delta_{\alpha,\beta} P_\alpha \quad \text{and} \quad \sum_\alpha P_\alpha = I. \tag{A.9}$$

Given that a normal operator $A : \mathcal{H} \to \mathcal{H}$ can be written as $A = UA_dU^\dagger$ with $A_d = \text{diag}\{a_1, a_2, \ldots, a_n\}$, a function $f : \mathcal{L}(\mathcal{H}) \to \mathbb{C}$

acting on A must be intended as

$$f(A) = Uf(A_d)U^\dagger, \tag{A.10}$$

with $f(A_d) := diag\{f(a_1), f(a_2), \ldots, f(a_n)\}$.

A.2.2 *Commutator and anti-commutator*

Definition A.2.9. The commutator (resp., anti-commutator) of two operators $A, B : \mathcal{H} \to \mathcal{H}$ is defined as $[A, B] := AB - BA$ (resp., $\{A, B\} := AB + BA$). When it is zero, we say that the operators commute (resp., anticommute).

Theorem A.2.10. *Two linear and Hermitian operators $A, B : \mathcal{H} \to \mathcal{H}$ commute if and only if they are simultaneously diagonalizable, i.e. there exists an orthonormal basis $\{|i\rangle\}_i$ such that both A and B are diagonal with respect to it and hence have spectral representations $A = \sum_i a_i|i\rangle\langle i|$, $B = \sum_i b_i|i\rangle\langle i|$.*

A.2.3 *Matrix decomposition*

The following decompositions are often useful to factorize matrices representing operators.

Theorem A.2.11 (Polar decomposition). *Given a square matrix A with complex entries, there exists a unitary matrix U and a positive semidefinite Hermitian matrix P such that*

$$A = UP.$$

This decomposition always exists; and as long as A is invertible, it is unique, with P positive-definite. The matrix P is always unique, even if A is singular, and given by $\sqrt{A^\dagger A}$. One can also decompose A in the form $A = KU$, where $K = \sqrt{AA^\dagger}$.

Theorem A.2.12 (Singular value decomposition). *Given an $m \times n$ matrix A with complex entries, there exists an $m \times m$ unitary* *matrix U and an $n \times n$ unitary matrix V such that*

$$A = UDV,$$

with D diagonal $(m \times n)$ matrix. Its entries are called the singular values of A.

A.3 Tensor product

Tensor product is a way to assemble together vector spaces, and it is closely connected to bilinear functions. Since our interest is primarily with Hilbert spaces, we can without loss of generality simply identify the notions of tensor and Kronecker products.

Definition A.3.1. Let $\mathcal{H}_1$ and $\mathcal{H}_2$ be Hilbert spaces with orthonormal basis $\{|j_1\rangle\}_{j_1=1}^m$ and $\{|j_2\rangle\}_{j_2=1}^n$, respectively. Their tensor product, denoted as $\mathcal{H}_1 \otimes \mathcal{H}_2$, is an $m \times n$-dimensional Hilbert space having as orthonormal basis the ordered pairs $(|j_1\rangle, |j_2\rangle)$. We also denote $(|j_1\rangle, |j_2\rangle) = |j_1\rangle \otimes |j_2\rangle$ and say that $|j_1\rangle \otimes |j_2\rangle$ is the tensor product of basis vectors $|j_1\rangle$ and $|j_2\rangle$.

Tensor products of other vectors than basis vectors are defined by requiring that the product be bilinear,

$$\left(\sum_{j_1=1}^{m} \alpha_{j_1}|j_1\rangle\right) \otimes \left(\sum_{j_2=1}^{n} \beta_{j_2}|j_2\rangle\right) = \sum_{j_1=1}^{m}\sum_{j_2=1}^{n} \alpha_{j_1}\beta_{j_2}|j_1\rangle \otimes |j_2\rangle. \tag{A.11}$$

Since $|j_1\rangle \otimes |j_2\rangle$ form the basis of $\mathcal{H}_1 \otimes \mathcal{H}_2$, the notion of the tensor product of vectors is perfectly well established, but note carefully that the tensor product is not commutative!

Often we use the shorthand notation $|uv\rangle$ or $|u\rangle|v\rangle$ in place of $|u\rangle \otimes |v\rangle$, also because it is closer to the original idea of regarding $|u\rangle \otimes |v\rangle$ as a pair $(|u\rangle, |v\rangle)$.

Suppose now $|u\rangle \in \mathcal{H}_1$, $|v\rangle \in \mathcal{H}_2$ and $A_1 : \mathcal{H}_1 \to \mathcal{H}_1$, $A_2 : \mathcal{H}_2 \to \mathcal{H}_2$, then we can define $A_1 \otimes A_2$ on $\mathcal{H}_1 \otimes \mathcal{H}_2$ as

$$(A_1 \otimes A_2)\left(|u\rangle \otimes |v\rangle\right) := \left(A_1|u\rangle\right) \otimes \left(A_2|v\rangle\right). \tag{A.12}$$

Next, since $\mathcal{H}_1 \otimes (\mathcal{H}_2 \otimes \mathcal{H}_3)$ is clearly isomorphic to $(\mathcal{H}_1 \otimes \mathcal{H}_2) \otimes \mathcal{H}_3$, we can omit the parenthesis and refer to this tensor product as $\mathcal{H}_1 \otimes \mathcal{H}_2 \otimes \mathcal{H}_3$. Thus, we can inductively define the tensor products of more than two spaces.

Whenever we have to assemble n identical spaces $\mathcal{H}$, we use the notation $\mathcal{H}^{\otimes n}$. Let $|\psi\rangle \in \mathcal{H}$ and $A : \mathcal{H} \to \mathcal{H}$, then

$$|\psi\rangle^{\otimes n} \equiv \underbrace{|\psi\rangle \otimes |\psi\rangle \otimes \cdots |\psi\rangle}_{n-\text{times}} \in \mathcal{H}^{\otimes n}, \tag{A.13}$$

$$A^{\otimes n} \equiv \underbrace{A \otimes A \otimes \cdots A}_{n-\text{times}} : \mathcal{H}^{\otimes n} \to \mathcal{H}^{\otimes n}. \tag{A.14}$$

Examples. In matrix representation, given two vectors

$$|\phi\rangle = \begin{pmatrix} \phi_1 \\ \phi_2 \\ \vdots \\ \phi_m \end{pmatrix}, \quad |\psi\rangle = \begin{pmatrix} \psi_1 \\ \psi_2 \\ \vdots \\ \psi_n \end{pmatrix}, \tag{A.15}$$

it will be

$$|\phi\rangle \otimes |\psi\rangle = \begin{pmatrix} \phi_1 \begin{pmatrix} \psi_1 \\ \psi_2 \\ \vdots \\ \psi_n \end{pmatrix} \\ \phi_2 \begin{pmatrix} \psi_1 \\ \psi_2 \\ \vdots \\ \psi_n \end{pmatrix} \\ \vdots \\ \phi_m \begin{pmatrix} \psi_1 \\ \psi_2 \\ \vdots \\ \psi_n \end{pmatrix} \end{pmatrix}. \tag{A.16}$$

Furthermore, if A (resp., B) is an $m \times m$ (resp., $n \times n$) matrix representing the operator A (resp., B) in some bases, then we have

$$A \otimes B \equiv \begin{pmatrix} A_{11}B & \cdots & A_{1n}B \\ \vdots & \ddots & \vdots \\ A_{m1}B & \cdots & A_{mn}B \end{pmatrix}, \tag{A.17}$$

where $A_{ij}B$ is the $n \times n$ submatrix with overall factor A_{ij}.

A.3.1 *Trace and partial trace*

Definition A.3.2. Given $A : \mathcal{H} \to \mathcal{H}$

$$\mathrm{Tr}\, A := \sum_{i=1}^{n} \langle e_i | A | e_i \rangle, \tag{A.18}$$

with $\{|e_i\rangle\}_{i=1}^{n}$ an orthonormal basis.

Note that the Trace is independent of the basis choice.

Furthermore, the Trace satisfies the following properties:

(i) $\mathrm{Tr}\,(aA + bB) = a\,\mathrm{Tr}\,A + b\,\mathrm{Tr}\,B$, for all A, B operators and $a, b \in \mathbb{C}$;
(ii) $\mathrm{Tr}\,A^\dagger = \mathrm{Tr}\,A$, for all A operators;
(iii) $\mathrm{Tr}\,(ABC) = \mathrm{Tr}\,(BCA) = \mathrm{Tr}\,(CAB)$, for all A, B, C operators.

Definition A.3.3. Let $X_{AB} : \mathcal{H}_A \otimes \mathcal{H}_B \to \mathcal{H}_A \otimes \mathcal{H}_B$ be a linear operator, then its partial trace on the space $\mathcal{H}_B$ is defined as

$$\mathrm{Tr}_B X_{AB} := \sum_i {}_B\langle e_i | X_{AB} | e_i \rangle_B,$$

where $\{|e_i\rangle_B\}_i$ is an orthonormal basis for $\mathcal{H}_B$.

Note that the partial trace gives an operator (in the above case acting on $\mathcal{H}_A$) not a c-number.

The trace on the global space $\mathcal{H}_A \otimes \mathcal{H}_B$ can always be decomposed into partial traces on the subsystems, i.e.

$$\mathrm{Tr}(X_{AB}) = \mathrm{Tr}_{AB}(X_{AB}) = \mathrm{Tr}_A\left(\mathrm{Tr}_B(X_{AB})\right) = \mathrm{Tr}_B\left(\mathrm{Tr}_A(X_{AB})\right). \tag{A.19}$$

A.4 Operator–vector correspondence

It is possible to establish a simple correspondence between the spaces $\mathfrak{L}(\mathcal{H})$ and $\mathcal{H} \otimes \mathcal{H}$ by defining a linear mapping that amounts to flipping a bra to a ket, i.e.

$$\mathsf{v}(|b\rangle\langle a|) := |b\rangle|a\rangle. \tag{A.20}$$

The v mapping is a linear bijection and as consequence it is clear that $\mathfrak{L}(\mathcal{H})$ is a vector space itself. Moreover, there we can define an inner product (denoted as $\langle\, ,\, \rangle$) of operators A, B as

$$\langle A, B\rangle = \langle \mathsf{v}(A)|\mathsf{v}(B)\rangle. \tag{A.21}$$

Thanks to Eq. (A.20), we have

$$A = \sum_{a,a'} \alpha_{a,a'}|a\rangle\langle a'| \xrightarrow{\mathsf{v}} \sum_{a,a'} \alpha_{a,a'}|a\rangle|a'\rangle \tag{A.22}$$

$$B = \sum_{b,b'} \beta_{b,b'}|b\rangle\langle b'| \xrightarrow{\mathsf{v}} \sum_{b,b'} \beta_{b,b'}|b\rangle|b'\rangle \tag{A.23}$$

Thus, it will be

$$\langle A, B\rangle = \langle \mathsf{v}(A)|\mathsf{v}(B)\rangle = \langle a|b\rangle\langle a'|b'\rangle = \sum_{a,a'} \alpha^*_{a,a'}\,\beta_{a,a'} = \mathrm{Tr}\left(A^\dagger B\right). \tag{A.24}$$

This is also know as Hilbert–Schmidt inner product and it makes the space $\mathfrak{L}(\mathcal{H})$ an Hilbert space itself. It also clearly induces a norm

$$||A||^2 = \langle A, A\rangle = \mathrm{Tr}\left(A^\dagger A\right). \tag{A.25}$$

However, other norms on $\mathfrak{L}(\mathcal{H})$ can be useful. In particular, the family of norms called *Schatten p-norms.*

Definition A.4.1. For any $A \in \mathfrak{L}(\mathcal{H})$ and any real number $p \geq 1$, the Schatten p-norm is defined as

$$||A||_p := \left\{\mathrm{Tr}\left[\left(A^\dagger A\right)^{p/2}\right]\right\}^{1/p}.$$

Remark A.4.2. The Schatten 1-norm is more commonly known as trace norm and corresponds to the one induced by the Hilbert–Schmidt inner product (A.25). The Schatten ∞-norm is the usual operator (spectral) norm. The latter coincides with $\max_v\{\|Av\| : v \in \mathcal{H}, \|v\| = 1\}$.

Here are some relevant properties satisfied by the Schatten p-norms:

(1) For any $A \in \mathfrak{L}(\mathcal{H})$ and every $p \in [1,\infty)$, it holds true that $\|A\|_p = \|A^\dagger\|_p$.
(2) For every $p \in [1,\infty)$, the Schatten p-norm is unitarily invariant, namely $\|A|_p = \|UAU^\dagger\|_p$ for any choice of unitary U on $\mathcal{H}$.
(3) The Schatten p-norms are non-increasing in p, i.e. $\|A\|_p \geq \|A\|_q$ for $1 \leq p \leq q < \infty$.
(4) For $p,q \in [1,\infty)$ such that $\frac{1}{p} + \frac{1}{q} = 1$, it holds $\|A\|_p = \max_B\{|\langle A,B\rangle| : B \in \mathfrak{L}(\mathcal{H}), \|B\|_q \leq 1\}$. This implies that

$$|\langle A,B\rangle| \leq \|A\|_p \, \|B\|_q, \tag{A.26}$$

which is the Holder inequality for Schatten norms.

A.5 Haar measure of U(d)

Since it holds $\mathrm{U}(d) = \mathrm{U}(1) \times \mathrm{SU}(d)$, let us focus on $\mathrm{SU}(d)$ starting from SU(2). Such a group, as a space, is homeomorphic to the sphere $\mathbb{S}^3$ on $\mathbb{R}^4$. To see this, it is enough to write a 2×2 matrix as

$$A = \begin{pmatrix} a & b \\ c & d \end{pmatrix},$$

with $a,b,c,d \in \mathbb{C}$. Such a matrix in order to belong to SU(2) must satisfy the conditions $A^\dagger A = I$ and $\det A = 1$, which imply $a = d^*$, $b = -d^*$ and $|a|^2 + |b|^2 = 1$. Then with the positions

$$x_1 = \frac{a+a^*}{2}, \quad x_2 = \frac{a-a^*}{2i}, \quad x_3 = \frac{b+b^*}{2}, \quad x_4 = \frac{b-b^*}{2i},$$

we can define a homeomorphism $A \leftrightarrow (x_1,x_2,x_3,x_4)$ between SU(2) and the sphere of unit radius in $\mathbb{R}^4$. Moving to the spherical coordinates, we set

$$x_1 = \cos\phi\cos\psi,$$
$$x_2 = \cos\phi\sin\psi,$$
$$x_3 = \sin\phi\cos\chi,$$
$$x_4 = \sin\phi\sin\chi,$$

with $0 \leq \phi \leq \pi/2$, $0 \leq \psi \leq 2\pi$ and $0 \leq \chi \leq 2\pi$. Then the volume element in terms of spherical coordinates results

$$\sqrt{2}\, d\left[(\sin\phi)^2\right] d\psi d\chi.$$

Hence, for $U \in \mathrm{U}(2)$, we can write

$$U = e^{i\alpha}\begin{pmatrix} e^{i\psi}\cos\phi & e^{i\chi}\sin\phi \\ -e^{-i\chi}\sin\phi & e^{-i\psi}\cos\phi \end{pmatrix},$$

with $e^{i\alpha}$ the U(1) factor, and

$$d\mu\,(U) = \frac{1}{4\pi^3}\, d\alpha\, \cos\phi\sin\phi\, d\phi\, d\psi\, d\chi, \qquad \text{(A.27)}$$

with $0 \leq \alpha \leq 2\pi$.

Now it is easy to see that for all $f \in L^1(\mathrm{U}(2))$, it holds

$$\int f(U'U)d\mu(U) = \int f(UU')d\mu(U) = \int f(U)d\mu(U),$$

simply by noting that

$$f(U'U) = f(\alpha' + \alpha, \phi' + \phi, \psi' + \psi, \chi' + \chi) = f(UU'),$$

and by performing the change of variables in the integral.

A measure like (A.27) satisfying such invariance property is said to be a *Haar measure.*

Quite generally, the group SU(d) for $d > 2$, as a space, is not a sphere. However, we can relate it to a sphere by the following well-known results:

$$\mathrm{SU}(d)/\mathrm{SU}(d-1) = \mathbb{S}^{2n-1}.$$

Thanks to that, it is possible to have a Haar measure for $\mathrm{U}(d)$ as

$$d\mu\,(U) = \frac{1}{2^d \pi^{d(d+1)/2}}\, d\alpha \prod_{1\leq k<\ell\leq d} \cos\phi_{k\ell}\,(\sin\phi_{k\ell})^{2k-1}\, d\phi_{k\ell}$$
$$\times\, d\psi_{k\ell} \prod_{1<\ell\leq d} d\chi_{\ell}.$$

Exercises

1. Prove that $|v_1\rangle\langle v_1| + |v_2\rangle\langle v_2| = |v_1'\rangle\langle v_1'| + |v_2'\rangle\langle v_2'|$, where

$$|v_1'\rangle = \cos\alpha|v_1\rangle - \sin\alpha|v_2\rangle,$$
$$|v_2'\rangle = \sin\alpha|v_1\rangle + \cos\alpha|v_2\rangle,$$

for any $\alpha \in \mathbb{R}$.

2. Show the following simple facts about linear operators:

(i) Any positive operator is Hermitian.
(ii) For a self-adjoint operator, the eigenvectors corresponding to distinct eigenvalues are orthogonal.
iii) The eigenvalues of a projector operator are either 0 or 1.

3. Prove Theorem A.2.10.
4. Using the polar decomposition, show that a normal operator $A : \mathcal{H} \to \mathcal{H}$ possesses the spectral decomposition.
5. Show the following simple facts about tensor product:

(i) The transposition, complex conjugation and adjointness distribute over the tensor product, that is for example $(A \otimes B)^\top = A^\top \otimes B^\top$.
(ii) The tensor product of two unitaries, Hermitians and projectors, is a unitary, Hermitian and projector, respectively.

6. Show that the trace function is independent of the chosen orthonormal basis.

7. Show that for any operator A, the operator $(A^\dagger A)^{1/2}$ is self-adjoint. Then, let $A = \begin{pmatrix} 1 & 1 \\ 0 & 1 \end{pmatrix}$, hence find $\|A\|_1$, $\|A\|_2$ and $\|A\|_{10}$.
8. Use the Holder inequality to show that the triangle inequality holds true for Schatten p-norms with $1 < p < \infty$. Further, show that the triangle inequality holds true also for $p = 1$.

SOLUTIONS TO SELECTED EXERCISES

Chapter 2

3. By the chain rule for entropy, it holds

$$H(X_1, X_2, \ldots, X_n) \leq \sum_{i=1}^{n} H(X_i|X_i - 1, \ldots, X_1)$$
$$\leq \sum_{i=1}^{n} H(X_i),$$

where the last inequality follows from the property of "conditioning reduces entropy". We have equality iff X_i is independent of $X_i - 1, \ldots, X_1$, for all i, i.e. if and only if X_is are independent.

4. On the one hand,

$$H(X, f(X)) = H(X) + H(f(X)|X)$$
$$= H(X),$$

because learning the value taken by X implies to learn the value taken by $f(X)$. On the other hand,

$$H(X, f(X)) = H(f(X)) + H(X|f(X))$$
$$\geq H(f(X)),$$

because learning the value taken by $f(X)$ does not necessarily imply to learn the value taken by X (unless the function is one-to-one). Thus, $H(g(X)) \leq H(X)$.

Chapter 3

4. The state vectors $|\Phi_k\rangle$ similarly to Bell's states form a basis, hence any entangling unitary can be written as

$$U = \sum_k e^{-i\lambda_k} |\Phi_k\rangle\langle\Phi_k|.$$

It is easy to show that

$$\begin{aligned}
U &= \exp\left[-\frac{i}{2}\left(\alpha_x \sigma^X \otimes \sigma^X + \alpha_y \sigma^Y \otimes \sigma^Y + \alpha_z \sigma^Z \otimes \sigma^Z\right)\right] \\
&= \exp\left[-\frac{i}{2}\alpha_x \sigma^X \otimes \sigma^X\right] \\
&\quad \times \exp\left[-\frac{i}{2}\alpha_y \sigma^Y \otimes \sigma^Y\right] \exp\left[-\frac{i}{2}\alpha_z \sigma^Z \otimes \sigma^Z\right] \\
&= \left[\cos\left(\frac{\alpha_x}{2}\right) I \otimes I - i \sin\left(\frac{\alpha_x}{2}\right) \sigma^X \otimes \sigma^X\right] \\
&\quad \times \left[\cos\left(\frac{\alpha_y}{2}\right) I \otimes I - i \sin\left(\frac{\alpha_y}{2}\right) \sigma^Y \otimes \sigma^Y\right] \\
&\quad \times \left[\cos\left(\frac{\alpha_z}{2}\right) I \otimes I - i \sin\left(\frac{\alpha_z}{2}\right) \sigma^Z \otimes \sigma^Z\right].
\end{aligned}$$

Then writing $I \otimes I$, $\sigma^X \otimes \sigma^X$, $\sigma^Y \otimes \sigma^Y$, $\sigma^Z \otimes \sigma^Z$ in terms of outer products in the computational basis and equating with the expression in terms of $|\Phi_k\rangle\langle\Phi_k|$, it is possible to find

$$\lambda_1 = \frac{\alpha_x - \alpha_y + \alpha_z}{2}, \quad \lambda_2 = \frac{-\alpha_x + \alpha_y + \alpha_z}{2},$$
$$\lambda_3 = \frac{-\alpha_x - \alpha_y - \alpha_z}{2}, \quad \lambda_4 = \frac{\alpha_x + \alpha_y - \alpha_z}{2}.$$

Chapter 4

5. It holds $p_e = p_0 \mathrm{Tr}(\rho_0 E_1) + p_1 \mathrm{Tr}(\rho_1 E_0)$. Since $E_0 + E_1 = I$, the probability of error can be rewritten as

$$p_e = p_0 + \mathrm{Tr}[(p_1\rho_1 - p_0\rho_0)E_0] = p_0 + \mathrm{Tr}(\Upsilon E_0),$$

where $\Upsilon := (p_1\rho_1 - p_0\rho_0)$. The problem is thus of finding the minimum of $\mathrm{Tr}(\Upsilon E_0)$ overall Hermitian operators $0 \leq E_0 \leq I$.

Writing $\Upsilon = \sum_j v_j |j\rangle\langle j|$, it is easy to see that the minimization is achieved by

$$E_0 = \sum_{j:v_j<0} |j\rangle\langle j|$$

and hence

$$p_e^{min} = p_0 + \sum_{j:v_j<0} v_j.$$

Chapter 5

4. Consider a purification of the state ρ_{12} using a Hilbert space $\mathcal{H}_{123}$. From the subadditivity property, we know that $S_3 + S_1 \geq S_{13}$. By construction $S_{123} = 0$, so that $S_{13} = S_2$ and $S_3 = S_{12}$. By rearranging the terms, the desired inequality follows.
5. Since $\rho = |\psi\rangle\langle\psi|$ is a pure state, we have $F(\rho, \rho') = |\langle\psi|\rho'|\psi\rangle|$, hence $d(\rho, \rho') = 2 - 2|\langle\psi|\rho'|\psi\rangle|$. We can argue that ρ' minimizing the distance should live in the subspace spanned by $|00\rangle$ and $|11\rangle$, and being separable must be of the form $\omega := a^2|00\rangle\langle 00| + (1 - a^2)|11\rangle\langle 11|$. Then let us write

$$\rho' = \lambda\varpi + (1-\lambda)\tau, \quad 0 \leq \lambda \leq 1,$$

with

$$\varpi := (|\alpha|^2|00\rangle\langle 00| + |\beta|^2|11\rangle\langle 11|)$$

and τ a generic density operator. Note that in the first term we have replaced ω by ϖ since the difference between them can always be incorporated into τ. Then

$$\begin{aligned}\frac{dd(\rho, \rho')}{d\lambda} &= -2(|\alpha|^4 + |\beta|^4 - \langle\psi|\tau|\psi\rangle) \\ &= -2(-2|\alpha|^2|\beta|^2 + 1 - \langle\psi|\tau|\psi\rangle).\end{aligned}$$

If τ is close to ρ, we have $1 - \langle\psi|\tau|\psi\rangle \leq 2|\alpha|^2|\beta|^2$ and d is increasing around ϖ, where it takes the minimum value $4|\alpha|^2(1-|\alpha|^2)$.

Chapter 6

3. It can be easily checked that for any ρ such that supp $\rho \in \text{span}\{|1\rangle, |2\rangle\}$, it holds $\mathcal{N} \circ \mathcal{E}\rho = \rho$. It then follows that $I_{\text{coh}}(\rho, \mathcal{N} \circ \mathcal{E}) = S(\rho)$. Furthermore, it results $I_{\text{coh}}(\mathcal{E}(\rho), \mathcal{N}) = 2S(\rho) - 1$. Thus, there exist states ρ such that $I_{\text{coh}}(\rho, \mathcal{N} \circ \mathcal{E}) \leq I_{\text{coh}}(\mathcal{E}(\rho), \mathcal{N})$.
5. (i) The question is to find in general the four-qubit state vectors that satisfy the requirement that the state of every qubit pair is a mixture of the four Bell states.

   Making use of the Schmidt decomposition of $|\Psi\rangle_{RABC}$ for the bipartite partition RA vs BC, we realize that this state vector can be written as a superposition of double Bell state vectors

$$|\Psi\rangle_{RA;BC} = \{v|\Phi^+\rangle|\Phi^+\rangle + z|\Phi^-\rangle|\Phi^-\rangle + x|\Psi^+\rangle|\Psi^+\rangle + y|\Psi^-\rangle|\Psi^-\rangle\}_{RA;BC},$$

   where x, y, z and v are complex amplitudes (with $|x|^2 + |y|^2 + |z|^2 + |v|^2 = 1$).

   A remarkable feature of these double Bell state vectors is that they transform into superpositions of double Bell state vectors for the two other possible partitions of the four qubits $RABC$ into two pairs (RB vs AC, RC vs AB).

   Then, for the partition RB vs AC, we obtain

$$|\Psi\rangle_{RB;AC} = \{v'|\Phi^+\rangle|\Phi^+\rangle + z'|\Phi^-\rangle|\Phi^-\rangle + x'|\Psi^+\rangle|\Psi^+\rangle + y'|\Psi^-\rangle|\Psi^-\rangle\}_{RB;AC},$$

   with

$$v' = (v + z + x + y)/2,$$
$$z' = (v + z - x - y)/2r, \tag{15.1}$$

$$x' = (v - z + x - y)/2,$$
$$y' = (v - z - x + y)/2.$$

Similarly, for the partition RC vs AB,

$$|\Psi\rangle_{RC;AB} = \{v''|\Phi^+\rangle|\Phi^+\rangle + z''|\Phi^-\rangle|\Phi^-\rangle + x''|\Psi^+\rangle|\Psi^+\rangle + y''|\Psi^-\rangle|\Psi^-\rangle\}_{RC;AB},$$

with

$$\begin{aligned} v'' &= (v + z + x - y)/2, \\ z'' &= (v + z - x + y)/2, \\ x'' &= (v - z + x + y)/2, \\ y'' &= (v - z - x - y)/2. \end{aligned} \tag{15.2}$$

Thus, Eqs. (15.1) and (15.2) relate the amplitudes of the double Bell state vectors for the three possible partitions of the four qubits into two pairs, and thereby specify the entire set of asymmetric Pauli cloning machines.

Let us rewrite the amplitudes of $|\psi\rangle_{RA;BC}$ as a two-dimensional discrete function $\alpha_{m,n}$ with $m, n = 0, 1$,

$$\alpha_{0,0} = v, \quad \alpha_{0,1} = z, \quad \alpha_{1,0} = x, \quad \alpha_{1,1} = y.$$

Analogously, $|\psi\rangle_{RB;AC}$ can be characterized by a two-dimensional function $\beta_{m,n}$ defined as

$$\beta_{0,0} = v', \quad \beta_{0,1} = z', \quad \beta_{1,0} = x', \quad \beta_{1,1} = y'.$$

Using this notation, it appears that the two functions are related by a two-dimensional discrete Fourier transform

$$\beta_{m,n} = \frac{1}{2}\sum_{x=0}^{1}\sum_{y=0}^{1}(-1)^{nx+my}\alpha_{x,y}.$$

This reflects the fact that if output A is close to perfect ($\alpha_{m,n}$ is a peaked function), then output B is very noisy ($\beta_{m,n}$ is a flat function), and *vice versa*.

Thus, the transformation characterizing PCM can be written as

$$|\psi\rangle_X \to \sum_{m,n=0}^{1} \alpha_{m,n} U_{m,n} |\psi\rangle_A |B_{m,-n}\rangle_{B,C}, \tag{15.3}$$

where $|B_{m,n}\rangle_{B,C}$ are Bell state vectors

$$|B_{m,n}\rangle_{B,C} := \frac{1}{\sqrt{2}} \sum_{k=0}^{1} e^{2\pi i(kn/2)} |k\rangle_B |k \oplus m\rangle_C,$$

with $m, n = 0, 1$. The operators $U_{m,n}$ are defined as

$$U_{m,n} := \sum_{k=0}^{1} e^{2\pi i(kn/2)} |k \oplus m\rangle\langle k|,$$

and correspond to Pauli matrices: m labels the shift errors (the bit flip) while n labels the phase errors (phase flip).

(ii) Symmetric PCMs require that

$$|v'| = |v|, \quad |z'| = |z|, \quad |x'| = |x|, \quad |y'| = |y|,$$

which are satisfied by

$$v = x + y + z, \tag{15.4}$$

where x, y, z, and v can be assumed to be real.

The optimal symmetric PCM can be obtained by requiring that the two outputs A and B of a symmetric cloner are maximally independent. Using Eqs. (15.2) and (15.4), we obtain

$$v'' = x + z,$$
$$z'' = y + z,$$
$$x'' = x + y,$$
$$y'' = 0.$$

Therefore, we have

$$\rho_{RC} = \rho_{AB} = |x+z|^2|\Phi^+\rangle\langle\Phi^+| + |y+z|^2|\Phi^-\rangle\langle\Phi^-| \\ + |x+y|^2|\Psi^+\rangle\langle\Psi^+|.$$

Thus, we need to maximize the joint von Neumann entropy of the two outputs A and B,

$$S(A,B) = -\mathrm{Tr}(\rho_{AB}\log\rho_{AB}) = H[|x+z|^2, |y+z|^2, |x+y|^2].$$

It is easy to see that the solution with $x, y, z \geq 0$ that maximizes $S(A,B)$ is $x = y = z$, that is, the Pauli channel underlying outputs A and B reduces to a depolarizing channel. Actually, we get $x = y = z = 1/\sqrt{12}$, reflecting that A and B emerge both from a depolarizing channel

$$|\psi\rangle\langle\psi| \to \frac{2}{3}|\psi\rangle\langle\psi| + \frac{1}{3}(I/2).$$

The corresponding scaling factor is $2/3$, while the fidelity of cloning is $5/6$.

Chapter 7

3. Let us introduce the states $\tilde{\rho}_\theta := \lambda\sigma_\theta + (1-\lambda)\tau_\theta$. It is easy to show that $J_Q(\tilde{\rho}_\theta) = \lambda J_Q(\sigma_\theta) + (1-\lambda)J_Q(\tau_\theta)$. We now consider the partial trace on one qubit. Since $\mathrm{Tr}_{\mathbb{C}^2}$ is a stochastic (i.e. trace-preserving completely positive) map, the monotonicity of the SLD Fisher information with respect to a stochastic map shows that $J_Q(\tilde{\rho}_\theta) \geq J_Q(\mathrm{Tr}_{\mathbb{C}^2}\tilde{\rho}_\theta) = J_Q(\rho_\theta)$.
4. For a generic family of pure states, we have $\rho_\theta = |\psi_\theta\rangle\langle\psi_\theta|$. Since $\rho_\theta^2 = \rho_\theta$, we have

$$\frac{d\rho_\theta}{d\theta} = \frac{d\rho_\theta}{d\theta}\rho_\theta + \rho_\theta\frac{d\rho_\theta}{d\theta}$$

and thus

$$L_\theta = 2\frac{d\rho_\theta}{d\theta} = |\psi_\theta\rangle\langle d\psi_\theta/d\theta| + |d\psi_\theta/d\theta\rangle\langle\psi_\theta|.$$

Finally, we have

$$J_Q(\theta) = 4[\langle d\psi_\theta/d\theta | d\psi_\theta/d\theta\rangle + (\langle d\psi_\theta/d\theta|\psi_\theta\rangle)^2].$$

For the unitary family of pure states $|\psi_\theta\rangle = U_\theta|\psi\rangle$, we have

$$\begin{aligned} |d\psi_\theta/d\theta\rangle &= -iG|\psi_\theta\rangle, \\ \langle d\psi_\theta/d\theta|d\psi_\theta/d\theta\rangle &= \langle\psi|G^2|\psi\rangle, \\ \langle d\psi_\theta/d\theta|\psi_\theta\rangle &= -i\langle\psi|G|\psi\rangle. \end{aligned}$$

Thus,

$$J_Q(\theta) = 4\langle\psi|(G - \langle\psi|G|\psi\rangle)^2|\psi\rangle$$

and

$$\mathrm{Var}(\theta) \geq \frac{1}{4\langle(G - \langle G\rangle)^2\rangle}.$$

5. The minimum average cost is always attained by a separable probe state for $d > 2$.

Chapter 8

3. Rename $\rho^{\otimes n} = \rho$ and $\Pi(n, \epsilon) = \Pi$. Then

$$\begin{aligned} \|\rho - \Pi\rho\Pi\|_1 &= \|(I - \Pi + \Pi)\rho - \Pi\rho\Pi\|_1 \\ &\leq \|(I - \Pi)\rho\|_1 + \|\Pi\rho(I - \Pi)\|_1 \\ &= \mathrm{Tr}|(I - \Pi)\sqrt{\rho}\sqrt{\rho}| + \mathrm{Tr}|\Pi\sqrt{\rho}\sqrt{\rho}(I - \Pi)| \\ &\leq \sqrt{\mathrm{Tr}\{(I - \Pi)^2\rho\}\mathrm{Tr}\{\rho\}} \\ &\quad + \sqrt{\mathrm{Tr}\{\Pi\rho\}\mathrm{Tr}\{\rho(I - \Pi)^2\}} \\ &\leq \sqrt{\mathrm{Tr}\{(I - \Pi)\rho\}} + \sqrt{\mathrm{Tr}\{\rho(I - \Pi)\}} \\ &= 2\sqrt{\mathrm{Tr}\{(I - \Pi)\rho\}} \\ &\leq 2\sqrt{\epsilon}. \end{aligned}$$

The second inequality is obtained through the Cauchy–Schwarz inequality for the Hilbert–Schmidt norm.

4. The density operator $\rho = \frac{1}{2}(|+\rangle\langle+| + |-\rangle\langle-|)$ is diagonal in the $|a\rangle$, $|b\rangle$ basis and has eigenvalues 0.9 and 0.1, yielding an entropy of 0.469 qubit.

   For each three-qubit block, let Λ be the subspace spanned by vectors in the $|a\rangle$, $|b\rangle$ product basis which includes the majority of $|a\rangle$s,

$$\Lambda = span\{|aaa\rangle, |aab\rangle, |aba\rangle, |baa\rangle\}.$$

   Then the coding is as follows: first make a measurement to see if the state lies in the four-dimensional subspace Λ. If so, we can encode the state vector according to the following unitary.

$$|aaa\rangle \to |00\rangle$$
$$|aab\rangle \to |01\rangle$$
$$|aba\rangle \to |10\rangle$$
$$|baa\rangle \to |11\rangle.$$

   If the state vector is not found to be in Λ, we put the qubit into the state vector $|00\rangle$. The decoding is accomplished by inverting the unitary mapping, yielding a decoded signal in Λ.

   In this coding scheme, none of the actual three-qubit signal states are coded perfectly, since none of them lies in Λ. Let Π_Λ be the projector onto Λ, then the fidelity of the coding scheme is $\mathrm{Tr}(\rho^{\otimes 3}\Pi_\Lambda) = (0.9)^3 + 3(0.9)^2(0.1) = 0.972$.

   However, the actual fidelity is even higher because we have coded those signals not found in Λ into $|00\rangle$, which is then decoded as $|aaa\rangle$. That is, even the failures of the coding scheme have a substantial probability of passing a fidelity test measurement. The total fidelity maybe calculated to be

$$\mathrm{Tr}(\rho^{\otimes 3}\Pi_\Lambda) + \mathrm{Tr}(\rho^{\otimes 3}(I - \Pi_\Lambda))(0.9)^3 = 0.9924.$$

5. Suppose that Alice replaces the single-signal states by their purifications $P_i = |\psi_i\rangle\langle\psi_i|$ acting on the Hilbert space $\mathcal{H}_Q \otimes \mathcal{H}'$. As the source produces only two kinds of states, the entropy of

the ensemble of purifications can be calculated explicitly

$$S(\tilde{\rho}) = H_2\left[\frac{1}{2}(1+\sqrt{(p_1-p_2)^2+4p_1p_2|\langle\psi_1|\psi_2\rangle|^2})\right],$$

where H_2 is the binary entropy. The minimal entropy is obtained if the overlap of $|\psi_1\rangle$ and $|\psi_2\rangle$ is the largest. The supremum of the overlaps of purifications of states ρ_1 and ρ_2 is given by the fidelity

$$F(\rho_1,\rho_2) \equiv \max|\langle\psi_1|\psi_2\rangle|^2 = \left[\mathrm{Tr}\sqrt{\sqrt{\rho_1}\rho_2\sqrt{\rho_1}}\right]^2$$

so that we obtain

$$\mathtt{R} \leq S_{\min}(\tilde{\rho}) = H_2\left[\frac{1}{2}(1+\sqrt{(p_1-p_2)^2+4p_1p_2F(\rho_1,\rho_2)})\right].$$

It results $F = 7/8$ and hence $R \leq 0.2$.

Chapter 9

3. Alice sends the mixture of pure states

$$\rho = \sum_{i=1}^{4}\frac{1}{4}|x_i\rangle\langle x_i|,$$

whose matrix representation in the computational basis is

$$\rho = \begin{pmatrix} \frac{1}{2} & \sqrt{\frac{1}{24}}(1+e^{-2\pi i/3}+e^{-4\pi i/3}) \\ \sqrt{\frac{1}{24}}(1+e^{2\pi i/3}+e^{4\pi i/3}) & \frac{1}{2} \end{pmatrix}.$$

Its eigenvalues are $\{\frac{1}{2},\frac{1}{2}\}$, then $\chi(\rho) = 1$.

To achieve a mutual information greater than 0.4 bit, a suitable POVM has the following elements:

$$A_1 = \frac{1}{4\pi^2}\int_0^{2\pi} d\varphi \int_0^{\pi/2} d\theta\, |\theta,\varphi\rangle\langle\theta,\varphi|,$$

$$A_2 = \frac{1}{4\pi^2}\int_{-\pi/2}^{\pi/2} d\varphi \int_{\pi/2}^{\pi} d\theta\, |\theta,\varphi\rangle\langle\theta,\varphi|,$$

$$A_3 = \frac{1}{4\pi^2}\int_{\pi/2}^{\pi} d\varphi \int_{\pi/2}^{\pi} d\theta\, |\theta,\varphi\rangle\langle\theta,\varphi|,$$

$$A_1 = \frac{1}{4\pi^2}\int_{\pi}^{3\pi/2} d\varphi \int_{\pi/2}^{\pi} d\theta\, |\theta,\varphi\rangle\langle\theta,\varphi|,$$

with

$$|\theta,\varphi\rangle = \cos\frac{\theta}{2}|0\rangle + e^{i\varphi}\sin\frac{\theta}{2}|1\rangle.$$

Chapter 10

5. First note that the correctability condition can be written as

$$P_{\mathcal{C}} E_i^\dagger E_j P_{\mathcal{C}} = \mathsf{C}_{ij} P_{\mathcal{C}}, \tag{15.5}$$

with $P_{\mathcal{C}}$ projector onto the quantum code. Diagonalize the matrix C to obtain a diagonal matrix D and a new error basis J_i. The correctability condition then becomes

$$P_{\mathcal{C}} J_i^\dagger J_j P_{\mathcal{C}} = \mathsf{D}_{ij} P_{\mathcal{C}}. \tag{15.6}$$

Make use of the polar decomposition and of the correctability condition in Eq. (15.6) to obtain

$$J_i P_{\mathcal{C}} = U_i \sqrt{P_{\mathcal{C}} J_i^\dagger J_i P_{\mathcal{C}}} = \sqrt{\mathsf{D}_{ii}}\, U_i P_{\mathcal{C}},$$

where U is some unitary matrix. Thus, the action of the error J_i is to rotate $\mathcal{C}$ into the subspace defined by the projector $\Pi_i := U_i P_{\mathcal{C}} U_i^\dagger = J_i P_{\mathcal{C}} U_i^\dagger/\sqrt{\mathsf{D}_{ii}}$. Since such subspaces are orthogonal by Eq. (15.6) and are in number of $\mathrm{rank}(\mathsf{D}) = \mathrm{rank}(\mathsf{C})$, it follows that $d \geq |\mathcal{C}|\,\mathrm{rank}(\mathsf{C})$. Finally, the non-degeneracy hypothesis, i.e. non-singularity of C, means $\mathrm{rank}(\mathsf{C}) = \mathrm{rank}(\mathcal{E})$ and then the statement follows. Note that $\text{rank}(\mathcal{N})$ corresponds to the minimal number of Kraus operators for $\mathcal{N}$.

If we look at the case of qubits where the noise acts independently and wish to correct noise affecting at most t qubits, we consider a basis of error operators given by products of Pauli matrices involving up to t qubits. Then, correcting all errors is equivalent to correcting the random-unitary channel $\mathcal{N}$ whose Kraus operators are proportional to the possible products of $i \leq t$ Pauli matrices. Since Pauli matrices are orthogonal, this Kraus representation is already minimal, whence $\text{rank}(\mathcal{N})$ can be straightforwardly derived counting the number of independent Kraus operators (the ones affecting i qubits are $3^i\binom{n}{i}$) and this leads to

$$2^n \geq 2^k \sum_{i=0}^{t} 3^i \binom{n}{i}.$$

For the channel $\mathcal{N}$, the generalized Hamming bound reads

$$2^k \left[1 + 3\binom{n}{2}\right] \leq 2^n.$$

For $k = 1$, it gives $n = 7$, however, an example of code with $n < 7$ that works well is given by

$$|0_L\rangle = |000\rangle, \quad |1_L\rangle = |111\rangle.$$

These basis codewords are not affected by the action of σ^Z on any pair of qubits. Consequently, the action of σ^X on a pair of qubits is identical to the action of σ^Y on the same pair of qubits. In other words, the code is degenerate. Therefore, one has only to correct errors due to σ^X operators. This can be realized through a projective measurement onto the subspaces $\mathcal{S}_{00} = \text{span}\{|000\rangle, |111\rangle\}$, $\mathcal{S}_{01} = \text{span}\{|100\rangle, |011\rangle\}$, $\mathcal{S}_{10} = \text{span}\{|010\rangle, |101\rangle\}$ and $\mathcal{S}_{11} = \text{span}\{|001\rangle, |110\rangle\}$. If the measurement outcome is "00", no errors have affected the qubits; on the contrary, if the measurement outcome is "01", errors have affected qubits 2 and 3 and can be corrected by applying σ^X

there. Similarly, all the other possible errors can be detected and corrected.

Chapter 11

3. It holds $C(\mathcal{N}) = C_1(\mathcal{N}) = 1 - H_2\left(\frac{1+\xi}{2}\right)$, where ξ is the maximum among $|p_0+p_1-p_2-p_3|$, $|p_0-p_1+p_2-p_3|$ and $|p_0-p_1-p_2+p_3|$.
   For a depolarizing channel, for which $p_1 = p_2 = p_3 = p$ and $p_0 = 1 - 3p/4$, it holds $C_E(\mathcal{N}) = 2 + 3\left(\frac{2-p}{4}\right)\log\left(\frac{2-p}{4}\right) + \left(\frac{3p-2}{4}\right)\log\left(\frac{3p-2}{4}\right)$.
4. It holds $Q(\mathcal{N}) = \max_{q\in[0,1]}[H_2(\eta q) - H_2(q - \eta q)]$ for $\eta\in[1/2,1]$, $Q(\mathcal{N}) = 0$ otherwise.
5. Any $\mathcal{N}_1$ and $\mathcal{N}_2$ that are given by an anti-degradable channel and entanglement breaking channel, respectively, possess such a property (super-activation).

Chapter 12

3. Writing the state vectors in the Schmidt decomposition,

$$|\psi_1\rangle = \sum_{i=1}^{n} \sqrt{\alpha_i}|i\rangle_A|i\rangle_B,$$

$$|\psi_2\rangle = \sum_{i=1}^{m} \sqrt{\alpha_i'}|i\rangle_A|i\rangle_B,$$

with coefficients $\alpha_1 \geq \cdots \geq \alpha_n$ and $\alpha_1' \geq \cdots \geq \alpha_m'$ one can easily check that $\alpha_1 < \alpha_1'$, but $\alpha_1 + \alpha_2 > \alpha_1' + \alpha_2'$, thus $|\psi_1\rangle \not\leftrightarrow |\psi_2\rangle$.

In contrast, the Schmidt coefficients γ_i and γ_i' of the state vectors tensored with $|\phi\rangle$ (given in decreasing order) are

$$|\psi_1\rangle|\phi\rangle : 0.24,\ 0.24,\ 0.16,\ 0.16,\ 0.06,\ 0.06,\ 0.04,\ 0.04,$$

$$|\psi_2\rangle|\phi\rangle : 0.30,\ 0.20,\ 0.15,\ 0.15,\ 0.10,\ 0.10,\ 0.00,\ 0.00.$$

so that $\sum_{i=1}^{k}\gamma_i \leq \sum_{i=1}^{k}\gamma_i'$ for $1 \leq k \leq 8$, and consequently, $|\psi_1\rangle|\phi\rangle \leftrightarrow |\psi_2\rangle|\phi\rangle$.

Now suppose that $|\phi\rangle = \frac{1}{\sqrt{d}}\sum_{i=1}^{d}|i\rangle_A|i\rangle_B$. The Schmidt coefficients of $|\psi_1\rangle|\phi\rangle$ will be α_i/d, each one being d-fold degenerate. Then we can write, for any l, that $\sum_{j=1}^{dl}\gamma_j = \sum_{i=1}^{l}\alpha_i$. Now, if $|\psi_1\rangle \not\rightarrow |\psi_2\rangle$ under LOCC, then for some $l = l_0$, we must have

$$\sum_{i=1}^{l_0}\alpha_i > \sum_{i=1}^{l_0}\alpha_i' \Rightarrow \sum_{i=1}^{dl_0}\gamma_i > \sum_{i=1}^{dl_0}\gamma_i' \Rightarrow |\psi_1\rangle|\phi\rangle \not\rightarrow |\psi_2\rangle|\phi\rangle,$$

under LOCC.

4. For a pure state $\rho = |\psi\rangle\langle\psi|$, the concurrence becomes $C = \lambda_1 = |\langle\psi|\theta|\psi\rangle|$. Expressing the state as

$$|\psi\rangle = a_{00}|00\rangle + a_{01}|01\rangle + a_{10}|10\rangle + a_{11}|11\rangle,$$

it results

$$C = 2|a_{00}a_{11} - a_{01}a_{10}|,$$

while the reduced density matrix results

$$\begin{pmatrix} |a_{00}|^2 + |a_{01}|^2 & a_{00}\overline{a_{10}} + a_{01}\overline{a_{11}} \\ \overline{a_{00}}a_{10} + \overline{a_{01}}a_{11} & |a_{10}|^2 + |a_{11}|^2 \end{pmatrix}.$$

Chapter 13

4. Consider a PCM acting as

$$|\psi\rangle_A \rightarrow \sum_{m,n=0}^{1} a_{m,n}\, U_{m,n}|\psi\rangle_B|B_{m,-n}\rangle_{E,E'}, \tag{15.7}$$

where A, B, E and E' stand for Alice's qubit, Bob's clone, Eve's clone and cloning machine, respectively. Here, the amplitudes $a_{m,n}$ (with $\sum_{m,n=0}^{1}|a_{m,n}|^2 = 1$) characterize the cloner, while the state vectors $|B_{m,n}\rangle_{EE'}$ are Bell state vectors

$$|B_{m,n}\rangle_{EE'} := \frac{1}{\sqrt{2}} \sum_{k=0}^{1} e^{2\pi i(kn/2)} |k\rangle_E |k \oplus m\rangle_{E'},$$

with $m, n = 0, 1$. The operators $U_{m,n}$ are defined as

$$U_{m,n} := \sum_{k=0}^{1} e^{2\pi i(kn/2)} |k \oplus m\rangle\langle k|,$$

and correspond to Pauli matrices: m labels the shift errors (the bit flip), while n labels the phase errors (phase flip).

Tracing the output joint state given by (15.7) over EE' implies that Alice's state $|\psi\rangle_A\langle\psi|$ is transformed, at Bob's station, into the mixture

$$\rho_B = \sum_{m,n=0}^{1} |a_{m,n}|^2 U_{m,n} |\psi\rangle\langle\psi| U_{m,n}^{\dagger}. \qquad (15.8)$$

Thus, the state undergoes a $U_{m,n}$ error with probability $|a_{m,n}|^2$.

If Alice sends any state $|k\rangle\langle k|$, with $|k\rangle \in \{|0\rangle, |1\rangle\}$, the phase errors ($n \neq 0$) clearly do not play any role in the above mixture since $U_{m,n}|k\rangle = e^{2\pi i(kn/2)}|k \oplus m\rangle$, so Bob's fidelity can be expressed as

$$F = \langle k|\rho_B|k\rangle = \sum_{n=0}^{d-1} |a_{0,n}|^2.$$

In the complementary basis, denoted for the sake of simplicity as $\{\overline{|0\rangle} \equiv |+\rangle, \overline{|1\rangle} \equiv |-\rangle\}$, we have $U_{m,n}\overline{|l\rangle} = e^{-2\pi i(l+n)m/2}\,\overline{|l \oplus n\rangle}$, so the shift errors ($m \neq 0$) do not play any role and Bob's fidelity becomes

$$\overline{F} = \overline{\langle l|}\rho_B\overline{|l\rangle} = \sum_{m=0}^{1} |a_{m,0}|^2.$$

For the cloner to copy equally well the states of both bases, we choose a 2×2 amplitude matrix of the form

$$a = \begin{pmatrix} v & x \\ x & y \end{pmatrix},$$

with x, y and v being real variables satisfying the normalization condition $v^2 + 2x^2 + y^2 = 1$. Thus, Bob's fidelity is $F = v^2 + x^2$ in both bases, and the corresponding mutual information between Alice and Bob is given by

$$I(A:B) = \log 2 + F \log F + (1 - F) \log(1 - F). \qquad (15.9)$$

Now, the clone kept by Eve is in a state given by an expression like (15.8), but with the amplitudes $a_{m,n}$ replaced by

$$b_{m,n} = \frac{1}{2} \sum_{m',n'=0}^{1} e^{2\pi i(nm' - mn')/2} a_{m',n'}.$$

This corresponds to an amplitude matrix with

$$x \to x' = (v - y)/2,$$
$$y \to y' = (v - 2x + y)/2,$$
$$v \to v' = (v + 2x + y)/2,$$

resulting in a cloning fidelity for Eve, $F_E = v'^2 + x'^2$. Maximizing the latter for a given value of Bob's fidelity F (using the normalization relation) yields the optimal cloner

$$x = \sqrt{F(1 - F)}, \quad y = 1 - F, \quad v = F.$$

The corresponding optimal fidelity for Eve is

$$F_E = \frac{1}{2} + \sqrt{F(1 - F)}.$$

As a consequence, the mutual information between Alice and Eve is given by

$$I(A:E) = \log 2 + F_E \log F_E + (1 - F_E) \log(1 - F_E). \quad (15.10)$$

Finally, Bob's and Eve's information (15.9), (15.10) intersect exactly where the fidelities coincide, that is, at

$$F = F_E = \frac{1}{2}\left(1 + \frac{1}{\sqrt{2}}\right),$$

or for QBER $D = 1 - F$ at $D = \frac{\sqrt{2}-1}{2\sqrt{2}}$.

5. Let us start writing an inequality of the form

$$L(B, E) \leq H(B)_{\mathcal{E}} + H(E)_{\mathcal{E}} \leq U(B, E),$$

where $L(B, E)$ and $U(B, E)$ are unknown lower and upper bounds respectively. Given that

$$I(B : \mathcal{E}) = H(B)_{\mathcal{E}} - \sum_i p_i H(B)_{\rho_i},$$

we will have

$$I(B : \mathcal{E}) + I(E : \mathcal{E}) \leq U(B, E) - L(B, E).$$

Exploiting the fact that

$$H(B)_{\mathcal{E}} + H(E)_{\mathcal{E}} \leq 2 \log N \equiv U(B, E),$$

and

$$H(B)_{\mathcal{E}} + H(E)_{\mathcal{E}} \geq -2 \log c \equiv L(B, E),$$

we finally arrive at the desired relation

$$I(B : \mathcal{E}) + I(E : \mathcal{E}) \leq 2 \log(Nc). \qquad (15.11)$$

Now consider n runs of the BB84 protocol, hence $N = 2^n$. The ensemble $\mathcal{E}$ will be the one where Alice encoded the information. Let us relabel the bases used for each of the n qubits such that Alice uses n times the σ^X basis. Hence, Bob's observable is $(\sigma^X)^{\otimes n}$. By symmetry, Eve's optimal information about the correct bases is the same as her optimal information about the incorrect ones. Hence, one can bound her information by assuming she measures $(\sigma^Z)^{\otimes n}$. Accordingly, $c = 2^{-n/2}$ and (15.11) gives $H(A : B) + H(A : E) \leq n$. Next, using the requirement $H(A : B) > H(A : E)$, one deduces that a secret key can be established once $H(A : B) > n/2$. Finally, using the fact that $H(A : B) = n[1 - H_2(q)]$, one arrives at $q \leq 11\%$.

BIBLIOGRAPHY

The following are the textbooks and review articles that served as guidance for this book. In them, the interested readers can find, albeit in an indirect way, other references, including original research articles.

1. C. H. Bennett and P. W. Shor, Quantum information theory, *IEEE Transactions on Information Theory*, vol. 44, pp. 2724–2742 (1998).
2. F. Caruso, V. Giovannetti, C. Lupo and S. Mancini, Quantum channels and memory effects, *Reviews of Modern Physics*, vol. 86, pp. 1203–1259 (2014).
3. T. M. Cover and J. A. Thomas, *Elements of Information Theory*, Wiley & Sons (1991).
4. A. Galindo and M. A. Martín-Delgado, Information and computation: Classical and quantum aspects, *Reviews of Modern Physics*, vol. 74, pp. 347–423 (2002).
5. C. W. Helstrom, *Quantum Detection and Estimation Theory*, Academic Press (1976).
6. R. Hill, *A First Course in Coding Theory*, Oxford University Press (1986).
7. R. Horodecki, P. Horodecki, M. Horodecki and K. Horodecki, Quantum entanglement, *Reviews of Modern Physics*, vol. 81, pp. 865–890 (2009).
8. M. Keyl, Fundamentals of quantum information theory, *Physics Reports*, vol. 369, pp. 431–548 (2002).
9. K. Kraus, *States, Effects and Operations: Fundamental Notions of Quantum Theory*, Springer Verlag, Berlin Heidelberg (1983).
10. D. A. Lidar and T. A. Brun, *Quantum Error Correction*, Cambridge University Press, Cambridge (2013).

11. H.-K. Lo, T. Spiller and S. Popescu, *Introduction to Quantum Computation and Information*, (1999).
12. D. J. C. Mac Kay, *Information Theory, Inference, and Learning Algorithms*, Cambridge University Press, Cambridge (2003).
13. M. A. Nielsen and I. L. Chuang, *Quantum Computation and Quantum Information*, Cambridge University Press, Cambridge (2010).
14. A. Peres, *Quantum Theory: Concepts and Methods*, Kluwer Academic Press (1995).
15. C. E. Shannon, A mathematical theory of communication, *Bell System Technical Journal*, vol. 27, pp. 379–423 and 623–656 (1948).
16. C. E. Shannon and W. Weaver, *The Mathematical Theory of Communication*, University of Illinois Press (1949).
17. V. Vedral, The role of relative entropy in quantum information theory, *Reviews of Modern Physics*, vol. 74, pp. 197–233 (2002).
18. R. F. Werner, Quantum information theory — an invitation, *Springer Tracts in Modern Physics*, vol. 173, pp. 14–57 (2001).
19. M. M. Wilde, *Quantum Information Theory*, Cambridge University Press, Cambridge (2017).

EPILOGUE

While originally Shannon was motivated by problems of communication when he developed information theory, this latter then became a field on its own and here we have treated its quantum version as a further extension.

We left out the continuous alphabets (i.e. continuous variables) to avoid complications related to the choice of appropriate topology on the manifold of quantum states and to the unboundedness of operators in infinite-dimensional Hilbert spaces.

The treatment of each topic was based on consolidated results. There are however results at the forefront of current research and open problems that do not appear here. Just to mention a few:

- Besides i.i.d. maps that are simple enough to study, memory channels play an important role in practical applications. A plethora of models can be devised to account for memory effects, however, causal structure seems of predominant importance. Clear results are available only when the memory dies down over successive channel uses.
- Acquiring information about a physical entity might involve decisions, besides estimation. For instance, one must decide which of a set of statements, or hypothesis, best describes the system insofar as observations permit one to judge. Quantum hypothesis testing can be dealt with using one-shot versions of quantum entropies. The ϵ-smooth min- and max-entropies are formed by

taking these entropies and optimizing them over the set of states ϵ-close to the original.

- Several studies have been devoted to develop universal quantum data compression algorithms, i.e. suited not only for a specific source but working well enough for any source however, we are still far from having a quantum analogue to the celebrated Lempel–Ziv code.
- There is an ongoing effort in error correcting coding to develop codes whose rates are closer and closer to capacity, and this is of course true even in the quantum framework. Polar codes seem promising in this direction, as they have been generalized to quantum polar codes.
- Concerning data transmission, it is not clear if the regularized capacity formulae can be effectively computed or not. Perhaps there exist other quantities that better capture the information transmission capacities of quantum channels.
- We have seen that entanglement provides a fundamental resource for quantum information processing and its manipulation has been well formalized. It turns out that it is possible to develop a *resource theory* more general than the one including only entanglement. This has important implications for the emergence of a quantum formulation of thermodynamics in physics.
- For practical quantum cryptography, a further step to take is the device independent security. This is a security notion that does not rely on trusting that the quantum devices used are truthful. The analysis of such protocols has to consider scenarios of imperfect or even malicious devices.

Beyond its intrinsic appeal, quantum information theory offers many points of interactions with other disciplines, including probability theory (central limit theorem, large deviations, measure concentration, random processes), statistical inference (hypothesis testing, parameter estimation, pattern recognition and learning), computer science (computational complexity, data structures, cryptology), mathematics (functional and convex analysis, dynamical systems, combinatorics, number theory), physics (metrology, thermodynamics, statistical mechanics), economics (portfolio theory, econometrics) and biology (molecular biology, biological sensors).

As such, it opens up the perspective for novel theories with the potential for unexpected applications, such as quantum thermodynamics, quantum machine learning, quantum portfolio theory.

INDEX

Printed in the United States
By Bookmasters